DeWALT®

DATACOM

PROFESSIONAL REFERENCE

Paul Rosenberg

Created exclusively
for DeWALT by:

PAL publications®

www.palpublications.com
1-800-246-2175

Titles Available From DeWALT

DeWALT Trade Reference Series

Construction Professional Reference

Datacom Professional Reference

Electric Motor Professional Reference

Electrical Estimating Professional Reference

Electrical Professional Reference

HVAC Professional Reference

Lighting & Maintenance Professional Reference

Plumbing Professional Reference

Referencia profesional sobre la industria eléctrica

Security, Sound & Video Professional Reference

Wiring Diagrams Professional Reference

DeWALT Exam and Certification Series

Electrical Licensing Exam Guide

HVAC Technician Certification Exam Guide

This Book Belongs To:

Name:______________________________

Company: ___________________________

Title: ______________________________

Department: _________________________

Company Address: ____________________

____

____

Company Phone: ______________________

Home Phone: _________________________

Pal Publications, Inc.
374 Circle of Progress
Pottstown, PA 19464-3810

First edition published 2005

ISBN 0-9759709-3-3

09 08 07 06 05 5 4 3 2 1

Printed in the United States of America

A Note To Our Customers

We have manufactured this book to the highest quality standards possible. The cover is made of a flexible, durable and water-resistant material able to withstand the toughest on-the-job conditions. We also utilize the Otabind process which allows this book to lay flatter than traditional paperback books that tend to snap shut while in use.

Datacom Professional Reference is not a substitute for the National Electric Code®. National Electric Code® and NEC® are registered trademarks of the National Fire Protection Association, Inc. Quincy, MA.

Preface

In the past several years, the data and communications industries have become a very important force in modern life. At the same time, they have changed dramatically, and a great number of technicians, contractors, designers, sales people and suppliers have been drawn into this field. I think it is very important for such people to have an easy to use compendium of charts, tables, diagrams, graphs and terminology. This enables them to get quick answers without the difficulty of keeping several large books within reach.

We have utilized a design and typeface that will hold the most information per page while at the same time making a particular topic quick to reference.

Information covered in this manual is necessary for anyone in the field to have in one's possession at all times. Naturally, a topic may have been overlooked or not discussed in depth to suit all tradespeople. I will constantly monitor and update this book on a regular basis to not only include requested additional material, but to add new material from the ever growing amount of high technology as it develops.

Best wishes,

Paul Rosenberg

Chapter 1 – *Communications Electronics* 1-1

CHAPTER 1
COMMUNICATIONS ELECTRONICS

BASIC FORMULAS FOR ELECTRICITY

Ohms Law (AC Current):

For the following Ohms Law formulas, θ is the phase angle in degrees where current lags voltage (inductive circuit) or by which current leads voltage (capacitive circuit). In a resonant circuit the phase angle is 0° and Impedance = Resistance

Current in amps = Voltage in volts / Impedance in ohms

$$\text{Current in amps} = \sqrt{\frac{\text{Power in watts}}{\text{Impedance in ohms} \times \cos\theta}}$$

Current in amps = Power in watts / Voltage in volts x cos θ

Voltage in volts = Current in amps x Impedance in ohms

Voltage in volts = Power in watts / Current in amps x cos θ

$$\text{Voltage in volts} = \sqrt{\frac{\text{Power in watts} \times \text{Impedance in ohms}}{\cos\theta}}$$

Impedance in ohms = Voltage in volts / Current in amps

Impedance in ohms = Power in watts / (Current in amps^2 x cos θ)

Impedance in ohms = (Voltage in volts^2 x cos θ) / Power in watts

Power in watts = (Current in amps)^2 x Impedance in ohms x cos θ

Power in watts = Current in amps x Voltage in volts x cos θ

$$\text{Power in watts} = \frac{(\text{Voltage in volts})^2 \times \cos\theta}{\text{Impedance in ohms}}$$

BASIC FORMULAS FOR ELECTRICITY *(cont'd)*

Ohms Law (DC Current):

$$\text{Current in amps} = \frac{\text{Voltage in volts}}{\text{Resistance in ohms}} = \frac{\text{Power in watts}}{\text{Voltage in volts}}$$

$$\text{Current in amps} = \sqrt{\frac{\text{Power in watts}}{\text{Resistance in ohms}}}$$

Voltage in volts = Current in amps x Resistance in ohms

Voltage in volts = Power in watts / Current in amps

$$\text{Voltage in volts} = \sqrt{\text{Power in watts} \times \text{Resistance in ohms}}$$

Power in watts = $(\text{Current in amps})^2$ x Resistance in ohms

Power in watts = Voltage in volts x Current in amps

Power in watts = $(\text{Voltage in volts})^2$ / Resistance in ohms

Resistance in ohms = Voltage in volts / Current in amps

Resistance in ohms = Power in watts / $(\text{Current in amps})^2$

Resistors in Series (values in Ohms):

$$\text{Total Resistance} = \text{Resistance}_1 + \text{Resistance}_2 + \ldots\text{Resistance}_n$$

Two Resistors in Parallel (values in Ohms):

$$\text{Total Resistance} = \frac{\text{Resistance}_1 \times \text{Resistance}_2}{\text{Resistance}_1 + \text{Resistance}_2}$$

Multiple Resistors in Parallel (values in Ohms):

$$\text{Total Resistance} = \frac{1}{{}^1/\text{Resistance}_1 + {}^1/\text{Resistance}_2 + \ldots {}^1/\text{Resistance}_n}$$

BASIC FORMULAS FOR ELECTRICITY *(cont'd)*

Capacitors in Parallel (values in any farad):

Total Capacitance = $\text{Capacitance}_1 + \text{Capacitance}_2 + \ldots \text{Capacitance}_n$

Two Capacitors in Series (values in any farad):

$$\text{Total Capacitance} = \frac{\text{Capacitance}_1 \times \text{Capacitance}_2}{\text{Capacitance}_1 + \text{Capacitance}_2}$$

Multiple Capacitors in Series (values in any farad):

$$\text{Total Capacitance} = \frac{1}{{}^1/\text{Resistance}_1 + {}^1/\text{Resistance}_2 + \ldots {}^1/\text{Resistance}_n}$$

Quantity of Electricity in a Capacitor:

Q in coulombs = Capacitance in farads x Volts

Capacitance of a Capacitor:

Capacitance in picofarads =

$$0.0885 \times \frac{\text{Dielectric constant} \times \text{area in cm}^2 \times (\#\text{ of plates} - 1)}{\text{thickness of dielectric in cm}}$$

Self Inductance:

When including the effects of coupling, add 2 x mutual inductance if fields are adding and subtract 2 x mutual inductance if the fields are opposing. Examples below:

Series: $L_t = L_1 + L_2 + 2M$ or $L_t = L_1 + L_2 - 2M$

$$\text{Parallel: } L_T = \frac{1}{[(1/L_1 + M) + (1/L_2 + M)]}$$

BASIC FORMULAS FOR ELECTRICITY *(cont'd)*

Resonance:

Resonant frequency in hertz (where $X_L = X_C$) =

$$\frac{1}{2\pi\sqrt{\text{Inductance in henrys} \times \text{Capacitance in farads}}}$$

Reactance:

Reactance in ohms of an inductance is X_L
Reactance in ohms of a capacitance is X_C

$X_L = 2\pi FH$
$X_C = 1 \div 2\pi FC$
F = frequency, in Hertz, H = inductance, in henrys, and C = capacitance, in farads.

Impedance:

Impedance in ohms = (series) $\sqrt{\text{Resistance in ohms}^2 + (X_L - X_C)^2}$

Impedance in ohms = (parallel) $\dfrac{\text{Resistance in ohms} + \text{Reactance}}{\sqrt{\text{Resistance in ohms}^2 + \text{Reactance}^2}}$

Susceptance:

Susceptance in mhos = $\dfrac{\text{Reactance in ohms}}{\text{Resistance in ohms}^2 + \text{Reactance in ohms}^2}$

Admittance:

Admittance in mhos = $\dfrac{1}{\sqrt{\text{Resistance in ohms}^2 \times \text{Reactance in ohms}^2}}$

Admittance in mhos = 1 / Impedance in ohms

BASIC FORMULAS FOR ELECTRICITY *(cont'd)*

Power Factor:

Power Factor = cos (Phase Angle)

Power Factor = True Power / Apparent Power

Power Factor = Power in watts / volts x current in amps

Power Factor = Resistance in ohms / Impedance in ohms

Q or figure of Merit:

Q = Inductive Reactance in ohms / Series Resistance in ohms

Q = Capacitive Reactance in ohms / Series Resistance in ohms

Efficiency of any Device:

Efficiency = Output / Input

Sine Wave Voltage and Current:

Effective (RMS) value = 0.707 x Peak value

Effective (RMS) value = 1.11 x Average value

Average value = 0.637 x Peak value

Average value = 0.9 x Effective (RMS) value

Peak value = 1.414 x Effective (RMS) value

Peak value = 1.57 x Average value

Decibels:

db = 10 Log_{10} (power in Watts #1 / Power in Watts #2)

db = 10 Log_{10} (Power Ratio)

db = 20 Log_{10} (Volts or Amps #1 / Volts or Amps #2)

db = 20 Log_{10} (Volts or Current Ratio)

Power Ratio = $10^{(db/10)}$

Voltage or Current ratio = $10^{(db/10)}$

If impedances are not equal: db = 20 $\log_{10}$ $[(Volt_1 \sqrt{Z_2}) / (Volt_2 \sqrt{Z_1})]$

UNDERSTANDING DECIBELS

In data cabling, most energy and power levels, losses or attenuations are expressed in decibels rather than in Watts. The reason is simple: Transmission calculations and measurements are almost always made as *comparisons* against a reference: Received power compared to emitted power, energy in versus energy out (energy lost in a connection), etc.

- Generally, energy levels (emission, reception, etc.) are expressed in *dBm.* This signifies that the reference level of 0 dBm corresponds to 1 mW of power.

- Generally, power losses or gains (attenuation in a cable, loss in a connector, etc.) are expressed in dB.

- The unit dBμ is used for very low levels.

- Decibel measurement works as follows—a difference of 3 dB equals a doubling or halving of power.

- A 3 dB gain in power means that the optical power has been doubled. A 6 dB gain means that the power has been doubled, and doubled again, equaling four times the original power. A 3 dB loss of power means that the power has been cut in half. A 6 dB loss means that the power has been cut in half, then cut in half again, equaling one fourth of the original power.

- A loss of 3 dB in power is equivalent to a 50% loss. For example, 1 milliwatt of power in, and .5 milliwatt of power out.

- A 6 dB loss would equal a 75% loss. (1 milliwatt in, .25 milliwatt out.)

DECIBELS VS. VOLT AND POWER RATIOS

Voltage	Power	+ DB -	Power	Voltage
1.000	1.000	0.0	1.000	1.000
1.059	1.122	0.5	0.891	0.944
1.122	1.259	1.0	0.794	0.891
1.189	1.413	1.5	0.708	0.847
1.259	1.585	2.0	0.631	0.794
1.334	1.778	2.5	0.562	0.750
1.413	1.995	3.0	0.501	0.708
1.496	2.239	3.5	0.447	0.668
1.585	2.512	4.0	0.398	0.631
1.679	2.818	4.5	0.355	0.596
1.778	3.162	5.0	0.316	0.562
1.884	3.548	5.5	0.282	0.531
1.995	3.981	6.0	0.251	0.501
2.113	4.467	6.5	0.224	0.473
2.239	5.012	7.0	0.200	0.447
2.371	5.623	7.5	0.178	0.422
2.512	6.310	8.0	0.158	0.398
2.661	7.079	8.5	0.141	0.376
2.818	7.943	9.0	0.126	0.355
2.985	8.913	9.5	0.112	0.335
3.162	10.000	10.0	0.100	0.316
3.350	11.220	10.5	0.089	0.299
3.548	12.589	11.0	0.079	0.282
3.758	14.125	11.5	0.071	0.266
3.981	15.849	12.0	0.063	0.251
4.217	17.783	12.5	0.056	0.237
4.467	19.953	13.0	0.050	0.224
4.732	22.387	13.5	0.045	0.211
5.012	25.119	14.0	0.040	0.200
5.309	28.184	14.5	0.035	0.188
5.623	31.623	15.0	0.032	0.178
5.957	35.481	15.5	0.028	0.168
6.310	39.811	16.0	0.025	0.158
6.683	44.668	16.5	0.022	0.150
7.079	50.119	17.0	0.020	0.141
7.499	56.234	17.5	0.018	0.133
7.943	63.096	18.0	0.016	0.126
8.414	70.795	18.5	0.014	0.119
8.913	79.433	19.0	0.013	0.112
9.441	89.125	19.5	0.011	0.106
10.0	100	20.0	0.010	0.100
31.6	1000	30.0	0.001	0.0316
100.0	10000	40.0	0.0001	0.01
316.2	10^5	50.0	0.00001	0.00316
1000	10^6	60.0	10^{-6}	0.001
3162	10^7	70.0	10^{-7}	0.000316
10000	10^8	80.0	10^{-8}	0.001
31620	10^9	90.0	10^{-9}	0.0000316
10^5	10^{10}	100.0	10^{-10}	10^{-5}
316200	10^{11}	110.0	10^{-11}	0.00000316
10^6	10^{12}	120.0	10^{-12}	10^{-6}

FORMULAS FOR COMMUNICATIONS

Frequency and Wavelength:

Frequency in kilohertz = (300,000) / wavelength in meters
Frequency in megahertz = (300) / wavelength in meters
Frequency in megahertz = (984) / wavelength in feet
Wavelength in meters = (300,000) / frequency in kilohertz
Wavelength in meters = (300) / frequency in megahertz
Wavelength in meters = (984) / frequency in megahertz

Length of an Antenna:

Quarter-wave antenna:
Length in feet = 234 / frequency in megahertz

Half-wave antenna:
Length in feet = 468 / frequency in megahertz

LCR Series Time Circuits:

Time in seconds =
Inductance in henrys / Resistance in ohms

Time in seconds =
Capacitance in farads x Resistance in ohms

70 Volt Loud Speaker Matching Transformer:

Transformer Primary Impedance =
(Amplifier output volts)^2 / Speaker Power

Time duration of One Cycle:

100 kilohertz	=	10 microsecond cycle
250 kilohertz	=	4 microsecond cycle
1 megahertz	=	1 microsecond cycle
4 megahertz	=	250 nanoseconds cycle
10 megahertz	=	100 nanoseconds cycle

COLOR CODES FOR RESISTORS

Radial Lead Resistor

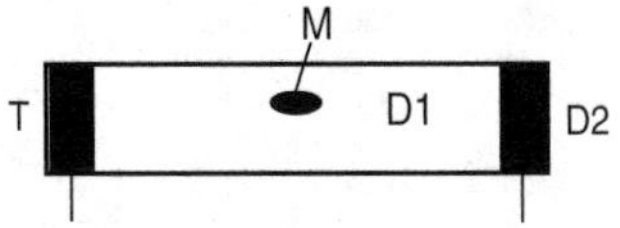

Axial Lead Resistor

Color	1st # (D1)	2nd # (D2)	Multiplier (M)	Tolerance (T)
No Color				20%
Silver			0.01	10%
Gold			0.1	5%
White	9	9	10^9	
Gray	8	8	100,000,000	
Violet	7	7	10,000,000	
Blue	6	6	1,000,000	
Green	5	5	100,000	
Yellow	4	4	10,000	
Orange	3	3	1,000	4%
Red	2	2	100	3%
Brown	1	1	10	2%
Black	0	0	1	1%

Example: Yellow – Blue – Brown – Silver = 460 ohms

If only Band D1 is wide, it indicates that the resistor is wirewound. If Band D1 is wide and there is also a blue fifth band to the right of Band T on the Axial Lead Resistor, it indicates the resistor is wirewound and flame proof.

STANDARD VALUES FOR RESISTORS

Values for 5% class   k = kilohms = 1,000 ohms   m = megohms = 1,000,000 ohms

1	8.2	68	560	4.7k	39k	330k	2.7m
1.1	9.1	75	620	5.1k	43k	360k	3.0m
1.2	10	82	680	5.6k	47k	390k	3.3m
1.3	11	91	750	6.2k	51k	430k	3.6m
1.5	12	100	820	6.8k	56k	470k	3.9m
1.6	13	110	910	7.5k	62k	510k	4.3m
1.8	15	120	1.0k	8.2k	68k	560k	4.7m
2.0	16	130	1.1k	9.1k	75k	620k	5.1m
2.2	18	150	1.2k	10k	82k	680k	5.6m
2.4	20	160	1.3k	11k	91k	750k	6.2m
2.7	22	180	1.5k	12k	100k	820k	6.8m
3.0	24	200	1.6k	13k	110k	910k	7.5m
3.3	28	220	1.8k	15k	120k	1.0m	8.2m
3.6	30	240	2.0k	16k	130k	1.1m	9.1m
3.9	33	270	2.2k	18k	150k	1.2m	10.0m
4.3	36	300	2.4k	20k	160k	1.3m	
4.7	39	330	2.7k	22k	180k	1.5m	
5.1	43	360	3.0k	24k	200k	1.6m	
5.6	47	390	3.3k	27k	220k	1.8m	
6.2	51	430	3.6k	30k	240k	2.0m	
6.8	56	470	3.9k	33k	270k	2.2m	
7.5	62	510	4.3k	36k	300k	2.4m	

CAPACITORS

Mica Capacitor

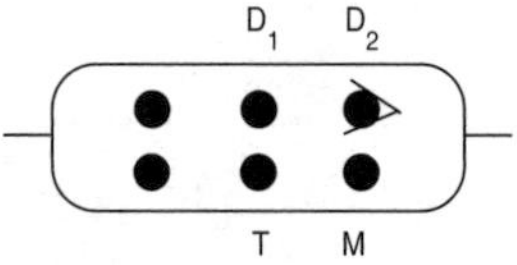

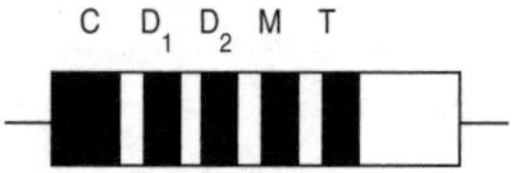

Ceramic Capacitor

Disc Capacitor

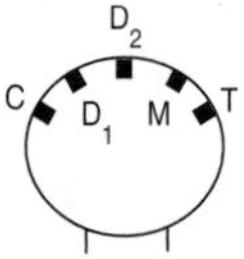

Ceramic disc capacitors are normally labeled. If the number is less than 1 then the value is in picofarads, if greater than 1 the value is in microfarads.The letter R can be used in place of a decimal; for example, 2R9=2.9

COLOR CODES FOR CAPACITORS

Color	1st # (D1)	2nd # (D2)	Multiplier (M)	Tolerance (T)
No Color				20%
Silver			0.01	10%
Gold			0.1	5%
White	9	9	10^9	9%
Gray	8	8	100,000,000	8%
Violet	7	7	10,000,000	7%
Blue	6	6	1,000,000	6%
Green	5	5	100,000	5%
Yellow	4	4	10,000	4%
Orange	3	3	1,000	3%
Red	2	2	100	2%
Brown	1	1	10	1%
Black	0	0	1	20%

COLOR CODES FOR CERAMIC CAPACITORS

Color	Tolerance (T) Above 10pf	Tolerance (T) Below 10pf	Decimal Multiplier (M)	Temp Coef ppm/°C (C)
White	10	1.0	0.1	500
Gray	–	0.25	0.01	30
Violet	–	–	–	-750
Blue	–	–	–	-470
Green	5	0.5	–	-330
Yellow	–	–	–	-220
Orange	–	–	1000	-150
Red	2	–	100	-80
Brown	1	–	10	-30
Black	20	2.0	1	0

STANDARD VALUES FOR CAPACITORS

pF	mF	mF	mF	mF
10	0.0010	0.10	10	1000
12	0.0012			
13	0.0013			
15	0.0015	0.15	15	
18	0.0018			
20	0.0020			
22	0.0022	0.22	22	2200
24				
27				
30				
33	0.0033	0.33	33	3300
36				
43				
47	0.0047	0.47	47	4700
51				
56				
62				
68	0.0068	0.68	68	6800
75				
82				
100	0.0100	1.00	100	10000
110				
120				
130				
150	0.0150	1.50		
180				
200				
220	0.0220	2.20	220	22000
240				
270				
300				
330	0.0330	3.30	330	
360				
390				
430				
470	0.0470	4.70	470	47000
510				
560				
620				
680	0.0680	6.80		
750				
820				82000
910				

pF = picofarads = 1 x 10^{-12} farads   mF = microfarads = 1 x 10^{-6} farads

CAPACITOR RATINGS

Typical Ratings*	Dimensions**		Model Number***
	Diameter	Length	
110-125 VAC, 50/60 Hz, Starting Capacitors			
88-106	1-7/16	2-3/4	EC8815
108-130	1-7/16	2-3/4	EC10815
130-489	1-7/16	2-3/4	EC13015
145-174	1-7/16	2-3/4	EC14515
161-193	1-7/16	2-3/4	EC16115
189-227	1-7/16	2-3/4	EC18915A
216-259	1-7/16	3-3/8	EC21615
233-280	1-7/16	3-3/8	EC23315A
243-292	1-7/16	3-3/8	EC24315A
270-324	1-7/16	3-3/8	EC27015A
324-389	1-7/16	3-3/8	EC2R10324N
340-408	1-13/16	3-3/8	EC34015
378-454	1-13/16	3-3/8	EC37815
400-480	1-13/16	3-3/8	EC40015
430-516	1-13/16	3-3/8	EC43015A
460-553	1-13/16	4-3/8	EC5R10460N
540-648	1-13/16	4-3/8	EC54015B
590-708	1-13/16	4-3/8	EC59015A
708-850	1-13/16	4-3/8	EC70815
815-978	1-13/16	4-3/8	EC81515
1000-1200	2-1/16	4-3/8	EC100015A
220-250 VAC, 50/60 Hz, Starting Capacitors			
53-64	1-7/16	3-3/8	EC5335
64-77	1-7/16	3-3/8	EC6435
88-406	1-13/16	3-3/8	EC8835
108-130	1-13/16	3-3/8	EC10835A
124-149	1-13/16	4-3/8	EC12435
130-154	1-13/16	4-3/8	EC13035
145-174	2-1/16	3-3/8	EC6R22145N
161-193	2-1/16	3-3/8	EC6R2216N
216-259	2-1/16	4-3/8	EC21635A
233-280	2-1/16	4-3/8	EC23335A
270-324	2-1/16	4-3/8	EC27035A

* in μF
** in inches
*** Model numbers vary by manufacturer

CAPACITOR RATINGS *(cont'd)*

270 VAC, 50/60 Hz, Running Capacitors

Typical Ratings*	Dimensions** Oval	Dimensions** Length	Model Number***
2		2-1/8	VH550
3		2-1/8	VH25503
4	1-5/16 x 2-5/32	2-1/8	VH5704
5		2-1/8	VH5705
6		2-5/8	VH5706
7.5		2-7/8	VH9001
10	1-5/16 x 2-5/32	2-7/8	VH9002
12.5		3-7/8	VH9003
15	1-29/32 x 2-29/32	3-1/8	VH9121
17.5		2-7/8	VH9123
20		2-7/8	VH5463
25	1-29/32 x 2-29/32	3-7/8	VH9069
30		3-7/8	VH5465
35	1-29/32 x 2-29/32	3-7/8	VH9071
40		3-7/8	VH9073
45	1-31/32 x 3-21/32	3-7/8	VH9115
50		3-7/8	VH9075

40 VAC, 50/60 Hz, Running Capacitors

Typical Ratings*	Dimensions** Oval	Dimensions** Length	Model Number***
10	1-5/16 x 2-5/32	3-7/8	VH5300
15	1-29/32 x 2-29/32	2-7/8	VH5304
17.5	1-29/32 x 2-29/32	3-7/8	VH9141
20	1-29/32 x 2-29/32	3-7/8	VH9082
25	1-29/32 x 2-29/32	3-7/8	VH5310
30		4-3/4	VH9086
35	1-29/32 x 2-29/32	4-3/4	VH9088
40		4-3/4	VH9641
45		3-7/8	VH5351
50	1-31/32 x 3-21/32	3-7/8	VH5320
55		4-3/4	VH9081

* in μF
** in inches
*** Model numbers vary by manufacturer

STANDARD WIRING COLOR CODES

Electronic applications (as established by the Electronic Industries Association – EIA):

Insulation Color	Circuit type
Black	Chassis grounds, returns, primary leads
Blue	Plate leads, transistor collectors, FET drain
Brown	Filaments, plate start lead
Gray	AC main power leads
Green	Transistor base, finish grid, diodes, FET gate
Orange	Transistor base 2, screen grid
Red	B plus dc power supply
Violet	Power supply minus
White	B – C minus of bias supply, AVC – AGC return
Yellow	Emitters-cathode and transistor, FET source

Stereo Audio Channels :

Insulation Color	Circuit type
White	Left channel high side
Blue	Left channel low side
Red	Right channel high side
Green	Right channel low side

AF Transformers (audio):

Insulation Color	Circuit type
Black	Ground line
Blue	Plate, collector, or drain lead. End of primary winding.
Brown	Start primary loop. Opposite to blue lead.
Green	High side, end secondary loop.
Red	B plus, center tap push - pull loop.
Yellow	Secondary center tap.

IF Transformers (Intermediate Frequency):

Insulation Color	Circuit type
Blue	Primary high side of plate, collector, or drain lead.
Green	Secondary high side for output.
Red	Low side of primary returning B plus.
Violet	Secondary outputs.
White	Secondary low side.

WAVELENGTHS – ELECTROMAGNETIC	
DESCRIPTION (BAND)	**FREQUENCY**
Non-visible Light Spectrum (Range)	.0005 Angstrom Units to .39 Micrometers (µM)
Cosmic Rays Gamma Rays X Rays Ultraviolet Light	.0005 Angstrom Units .0000006 to .00001 (µM) .00001 to .032 (µM) .032 to .39 (µM)
Visible Light Spectrum	.39 to 76 Micrometers (µM)
Violet Light Blue Light Green Light Maximum Visibility Yellow Light Orange Light Red Light	4240 to 4000 Angstrom Units 4912 to 4240 Angstrom Units 5960 to 4912 Angstrom Units 5750 to 5560 Angstrom Units 5850 to 5750 Angstrom Units 6740 to 5850 Angstrom Units 7000 to 6470 Angstrom Units
Infrared Heat & Light	.76 to 30 Micrometers (µM)
Extra-High Frequencies such as Gov't Experiments and Radar	300 Gigahertz (GHZ) to 30,000 - 3,000 Megahertz (MHZ)
Super High Frequencies such as Radio Navigation and Amateur Radio	30,000 to 3,000 Megahertz (MHZ)
Ultra High Frequencies	3000 to 300 Megahertz (MHZ)
Radars	1600 to 1300 (MHZ)
Non-Government and Government Radio Navigation, Amateur Radio Cellular Telephone Channels 14-69 (TV) Class A Citizens .7 Meter Amateur Radio	3000 to 890 (MHZ) 890 to 806 (MHZ) 806 to 470 (MHZ) 563.2 to 462.55 (MHZ) 420 to 400 (MHZ)
Very High Frequencies	300 to 30 Megahertz (MHZ)
United States Military 1-1/4 Meter Amateur Radio CB Radio, Navigation (AIR) Amateur Radio, Gov't & Non-Gov't Channels 7 – 13 (TV) 2 Meter Amateur Radio	400 to 275 (MHZ) 225 to 220 (MHZ) 470 to 216 (MHZ) 216 to 174 (MHZ) 148 to 144 (MHZ)

WAVELENGTHS – ELECTROMAGNETIC *(cont'd)*

DESCRIPTION (BAND)	FREQUENCY
United States Government	174 to 148 (MHZ)
Communication (Civil)	136 to 118 (MHZ)
Navigation (Aeronautical)	118 to 108 (MHZ)
FM Broadcast	108 to 88 (MHZ)
Channels 5 – 6 (TV)	88 to 76 (MHZ)
U.S. Gov't (Aeronautical)	76 to 72 (MHZ)
Channels 2 – 4 (TV)	72 to 54 (MHZ)
6 Meter Amateur Radio	54 to 50 (MHZ)
Railroads, Fire and Police	50 to 30 (MHZ)
High Frequencies	30 to 3 Megahertz (MHZ)
Class D Citizens	27.405 to 26.965 (MHZ)
Science, Medicine and Industries	27.54 to 26.95 (MHZ)
10 Meter Amateur Radio	29.7 to 28 (MHZ)
12 Meter Amateur Radio	24.99 to 24.89 (MHZ)
15 Meter Amateur Radio	21.45 to 21 (MHZ)
17 Meter Amateur Radio	18.168 to 18.068 (MHZ)
20 Meter Amateur Radio	14.35 to 14.1 (MHZ)
30 Meter Amateur Radio	10.15 to 10.1 (MHZ)
40 Meter Amateur Radio	7.3 to 7 (MHZ)
80 Meter Amateur Radio	4 to 3.5 (MHZ)
Medium Frequencies	3000 to 300 Kilohertz (KHZ)
AM Broadcast	1705 to 535 (KHZ)
160 Meter Amateur Radio	2000 to 1800 (KHZ)
Low Frequencies	300 to 30 Kilohertz (KHZ)
Communication/Navigation (Marine)	535 to 30 (KHZ)
All Radio Frequencies	30,000 MHZ to 30 (KHZ)
Very Low Frequency	30 KHZ to 3 (KHZ)
Ultrasonic Frequency	16 KHZ to 10 (KHZ)
Ultra Low Frequency	300 HZ to 30 HZ
Extremely Low Frequency	30 HZ to 3 HZ

WAVELENGTHS – MECHANICAL

DESCRIPTION (BAND)	FREQUENCY – HERTZ (HZ)
Normal Music	4186.01 to 16 HZ
Standard Range of Hearing (Human)	15000 to 30 HZ
Audio	20000 to 15 HZ
Direct Current (Continuous)	0 Hertz

SMALL TUBE FUSES

CHARACTERISTICS	Fuse Type	Fuse Dia.	Fuse Lgth.
Dual Element, Time Delay, glass tube	MDL	1/4"	1-1/4"
Dual Element, glass tube	MDX	1/4"	1-1/4"
Dual Element glass tube, Pigtail	MDV	1/4"	1-1/4"
Ceramic body, normal, 200% 15sec	3AB	1/4"	1-1/4"
Metric, fast blow, high int., 210% 30minutes	216	5mm	20mm
Glass, Metric, fast blow, 210% 30minutes	217	5mm	20mm
Glass, Metric, slow blow, 210% 2minutes	218	5mm	20mm
No Delay, Ceramic, 110% rating,	ABC		
Opens at 135% load in one hour		1/4"	1-1/4"
Fast Acting, glass tube, 110% rating,	AGC		
Opens at 135% load in one hour		1/4"	1-1/4"
Fast Acting, glass tube	AGX	1/4"	1"
No delay, 200% 15sec	BLF	13/32"	1-1/2"
No delay, military, 200% 15sec	BLN	13/32"	1-1/2"
Fast cleaning, 600V, 135% 1hr	BLS	13/32"	1-3/8"
Time delay, indicator pin, 135% 1hr	FLA	13/32"	1-1/2"
Dual element, delay, 200% 12sec	FLM	13/32"	1-1/2"
Dual element, delay, 500V, 200% 12sec	FLQ	13/32"	1-1/2"
Slow Blow time Delay	FNM	13/32"	1-1/4"
Slow Blow, Indicator, metal pin pops outs	FNA		
Indicating Blown, dual element		13/32"	1-1/2"
Rectifier Fuse, Fast, low let through	GBB	1/4"	1-1/4"
Indicator Fuse, Metal pin pops out	GLD		
Indicating Blown, 110% rating		1/4"	1-1/4"
Metric, fast acting	GGS	5mm	20mm
Fast, current limiting, 600V, 135% 1 hr	KLK	13/32"	1-1/2"
Fast, protect solid state, 250% 1 sec	KLW	13/32"	1-1/2"
Slow Blow, time delay	SC	13/32"	1-5/6" to
Size rejection also			2-1/4"
Slow blow, glass body, 200% 5sec	218000	0.197"	0.787"
Slow Blow, glass body, 200% 5sec	313000	1/4"	1-1/4"
Slow Blow, ceramic, 200% 5sec	326000	1/4"	1-1/4"
Auto Glass, fast blow, 200% 5sec	1AG	1/4"	5/8"
Auto Glass, fast blow, 200% 10sec	2AG	0.177"	0.57"
Auto Glass, fast blow, 200% 5sec	3AG	1/4"	1-1/4"
Auto Glass, fast blow, 200% 5sec	4AG	9/32"	1-1/4"
Auto Glass, fast blow, 200% 5sec	8AG	13/32"	1-1/2"
Auto Glass, fast blow, 200% 5sec	7AG	1/4"	7/8"
Auto Glass, fast blow, 200% 5sec	8AG	1/4"	1"
Auto Glass, fast blow, 200% 5sec	9AG	1/4"	1-7/16"

The percentage and time figures mean that a 135% overload will blow a KLK type fuse in 1 hour. (example)

FUEL CELLS AND BATTERIES

TYPE OF UNIT	(–) CATHODE	(+) ANODE	ACTUAL WORKING VOLTAGE (WV)	VOLTAGE (TV)	AMP.-HRS. PER KG.
Fuel Cells					
Hydrogen	O_2	H_2	0.7	1.23	26,000
Hydrazine	O_2	N_2H_4	0.7	1.5	2,100
Methanol	O_2	CH_2OH	0.9	1.3	1,400
Batteries					
Cadm-Air (RC)	O_2	Cd	0.8	1.2	475
Edison (RC)	NiO.	Fe	1.2	1.5	195
Hyd.Perox. (RC)	O_2	H_2	0.8	1.23	3,000
Lead-Acid (RC)	PbO_2	Pb	2.0	2.1	55
NiCad (RC)	NiO	Cd	1.2	1.35	165
Silver Cadm. (RC)	AgO	Cd	1.05	1.4	230
Ammonia	m-DNB	Mg	1.7	2.2	1,400
Cuprous chloride	CuCl	Mg	1.4	1.5	240
Leclance	MnO_2	Zn	1.2	1.6	230
Lithium High-Temp with fused salt	S	Li	1.8	2.1	685
Magnesium	MnO_2	Mg	1.5	2.0	270
Mercury	HgO	Zn	1.2	1.34	185
Mercad	HgO	Cd	0.85	0.9	165
MnO_2	MnO_2	Zn	1.15	1.5	230
Organic Cath.	mDNB	Mg	1.15	1.8	1,400
Silver Chloride	AgCl	Mg	1.5	1.6	170
Silver Oxide	AgO	Zn	1.5	1.85	285
Silver-Poly	Polyiodide	Ag	0.6	0.66	180
Sodium High-Temp with electrolyte	S	Na	1.8	2.2	1,150
Thermal	Fuel	Ca	2.6	2.8	240
Zinc-Air	O_2	Zn	1.1	1.6	815
Zinc-Nickel	Ni oxides	Zn	1.6	1.75	185
Zinc-Silver Oxide	AgO	Zn	1.5	1.85	285

(RC) Indicates the cell is secondary and can be recharged. (WV) Indicates the average voltage generated by a working cell. (TV) Indicates the theoretical voltage developed by the cell. Amp-hours per kG. is the theoretical capacity of the cell.

CHARACTERISTICS OF LEAD – ACID BATTERIES

TEMPERATURE (F) VERSUS BATTERY EFFICIENCY (%)

–20°	—	18%	20°	—	58%
–10°	—	33%	30°	—	64%
0°	—	40%	50°	—	82%
10°	—	50%	80°	—	100%

CHARGE	SPECIFIC GRAVITY OF ACID
Discharged	1.11 to 1.12
Very Low Capacity	1.13 to 1.15
25% of Capacity	1.15 to 1.17
50% of Capacity	1.20 to 1.22
75% of Capacity	1.24 to 1.26
100% of Capacity	1.26 to 1.28
Overcharged	1.30 to 1.32

MAGNETIC PERMEABILITY OF SOME COMMON MATERIALS

Substance	Permeability (approx.)
Aluminum	Slightly more than 1
Bismuth	Slightly less than 1
Cobalt	60-70
Ferrite	100-300
Free space	1
Iron	60-100
Iron, refined	3000-8000
Nickel	50-60
Permalloy	3000-30,000
Silver	Slightly less than 1
Steel	300-600
Super permalloys	100,000-1,000,000
Wax	Slightly less than 1
Wood, dry	Slightly less than 1

TRANSISTOR CIRCUIT ABBREVIATIONS

Quantity	Abbreviations
Base-emitter voltage	E_B, V_B, E_{BE}, V_{BE}
Collector-emitter voltage	E_C, V_C, E_{CE}, V_{CE}
Collector-base voltage	E_{BC}, V_{BC}, E_{CB} V_{CB}
Gate-source voltage	E_G, V_G, E_{GS}, V_{GS}
Drain-source voltage	E_D, V_D, E_{DS}, V_{DS}
Drain-gate voltage	E_{DG}, V_{DG}, E_{DG}, V_{DG}
Emitter current	I_E
Base current	I_B, I_{BE}, I_{EB}
Collector current	I_C, I_{CE}, I_{EC}
Source current	I_S
Gate current	I_G, I_{GS}, I_{SG}*
Drain current	I_D, I_{DS}, I_{SD}

*This is almost always significant.

RADIO FREQUENCY CLASSIFICATIONS

Classification	Abbreviation	Frequency range
Very Low Frequency	VLF	9 kHz and below
Low Frequency (Longwave)	LF	30 kHz - 300 kHz
Medium Frequency	MF	300 kHz - 2MHz
High Frequency (Shortwave)	HF	3 MHz - 30MHz
Very High Frequency	VHF	30 MHz - 300 MHz
Ultra High Frequency	UHF	300 MHz - 3 GHz
Microwaves		3 GHz and more

DATA MODULATION METHODS

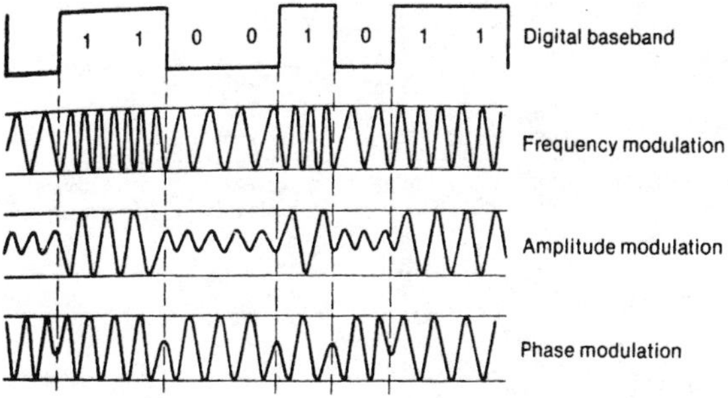

GENERAL BANDWIDTH REQUIREMENTS

Application	Bandwidth
Telephone-quality audio	64 kbits/sec
Simple application-sharing	100 kbits/sec
Videoconferencing	128 kbits/sec to 1 Mbit/sec
MPEG video	1.54 Mbits/sec
Imaging	8 to 100 Mbits/sec
Virtual reality	>100 Mbits/sec

TRANSMISSION MEDIA & CAPACITIES

Medium	Designation	Bit Rate (Mbps)	Voice Channels	Repeater Spacing (km)
Coaxial Cable	DS-1	1.544	24	1–2
	DS-1C	3.152	48	
	DS-2	6.312	96	
	DS-3	44.736	672	
Fiber (Sonet)	OC-1	51.84	672	25 (Laser)
	OC-3	155.52	2016	2 (LED)
	OC-9	466.56	6048	
	OC-12	622.08	8064	
	OC-18	933.12	12,096	
	OC-24	1244.16	16,128	
	OC-36	1866.24	24,192	
	OC-48	2488.32	32,256	
	OC-96	4976.64	64,512	
	OC-192	9953.28	129,024	

COMMON CARRIER AND OPERATIONAL FIXED (INDUSTRIAL) MICROWAVE FREQUENCY ALLOCATIONS IN THE UNITED STATES

Common Carrier	Operational Fixed
2.110 - 2.130 GHz	1.850 - 1.990 GHz
2.160 - 2.180 GHz	2.130 - 2.150 GHz
3.700 - 4.200 GHz	2.180 - 2.200 GHz
5.925 - 6.425 GHz	2.500 - 2.690 GHz (television)
10.7 - 11.700 GHz	6.575 - 6.875 GHz
	12.2 - 12.700 GHz*

*Based on noninterference with direct broadcast satellite service.

ELECTRONIC SYMBOLS

ammeter	
amplifier (operational)	
AND gate	
antenna (balanced, dipole)	
antenna (general)	
antenna (llop, shielded)	
antenna (unshielded)	
antenna (unbalanced)	
antenna (whip)	
attenuator (or resistor, fixed)	
attenuator (or resistor, variable)	
battery	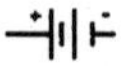

ELECTRONIC SYMBOLS *(cont'd)*

capacitor (feedthrough)	
capacitor (fixed, nonpolarized)	
capacitor (fixed, polarized)	
capacitor (ganged, variable)	
capacitor (single variable)	
capacitor (split-rotor, variable)	
capacitor (split-stator, variable)	
cathode (directly heated)	
cathode (indirectly heated)	
cathode (cold)	
cavity resonator	
cell	

ELECTRONIC SYMBOLS *(cont'd)*

coaxial cable	
coaxial cable (grounded shield)	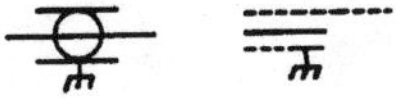
crystal (piezoelectric)	
delay line	
diode (field effect)	
diode (general)	
diode (Gunn)	
diode (light-emitting)	
diode (photosensitive)	
diode (photovoltaic)	
diode (pin)	
diode (Schottky)	

ELECTRONIC SYMBOLS *(cont'd)*

diode (tunnel)	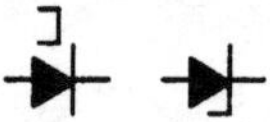
diode (varactor)	
diode (zener)	
directional coupler (or wattmeter)	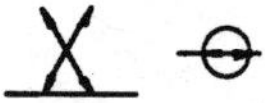
exclusive-OR gate	
female contact (general)	
ferrite bead	
fuse	
galvanometer	
ground (chassis)	
ground (earth)	
handset	

ELECTRONIC SYMBOLS *(cont'd)*

headphone (single)	
headphone (stereo)	
inductor (air-core)	
inductor (bifilar)	
inductor (iron-core)	
inductor (tapped)	
inductor (variable)	
integrated circuit	
inverter or inverting amplifier	
jack (coaxial or phono)	
jack (phone, two-conductor)	
jack (phone, two-conductor interrupting)	

jack (phone, three-conductor)

jack (phono)

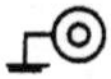

key (telegraph)

lamp (incandescent)

lamp (neon)

male contact (general)

meter (general)

microammeter

microphone

microphone (directional)

milliammeter

NAND gate

ELECTRONIC SYMBOLS *(cont'd)*

negative voltage connection 

NOR gate 

operational amplifier  

OR gate 

outlet (nonpolarized) 

outlet (polarized) 

outlet (utility, 117 V, nonpolarized) 

outlet (utility, 234 V) 

photocell (tube) 

plug (nonpolarized) 

plug (polarized) 

plug (phone, two-conductor) 

ELECTRONIC SYMBOLS *(cont'd)*

plug (phone, three-conductor) 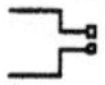

plug (phono) 

plug (utility, 117 V) 

plug (utility, 234V) 

positive-voltage connection

potentiometer (variable resistor, or rheostat) 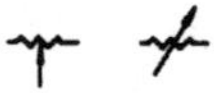

probe (radio-frequency) 

rectifier (semiconductor) 

rectifier (silicon-controlled) 

rectifier (tube-type) 

rectifier (tube-type, gas-filled) 

relay (DPDT) 

ELECTRONIC SYMBOLS *(cont'd)*

relay (DPST)	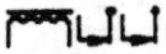
relay (SPDT)	
relay (SPST)	
resistor (fixed)	
resistor (preset)	
resistor (tapped)	
resonator	
rheostat (variable resistor, or potentiometer)	
saturable reactor	
shielding	
signal generator	
solar cell source (constant-voltage)	

ELECTRONIC SYMBOLS *(cont'd)*

<table>
<tr><td>source (constant-current)</td><td></td></tr>
<tr><td>speaker</td><td></td></tr>
<tr><td>switch (DPDT)</td><td>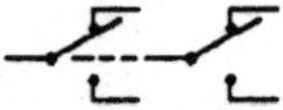</td></tr>
<tr><td>switch (DPST)</td><td></td></tr>
<tr><td>switch (momentary-contact)</td><td>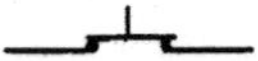</td></tr>
<tr><td>switch (rotary)</td><td>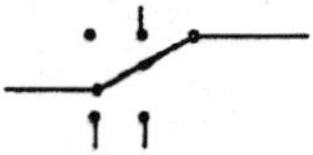</td></tr>
<tr><td>switch (silicon-controlled)</td><td></td></tr>
<tr><td>switch (SPDT)</td><td></td></tr>
<tr><td>switch (SPST)</td><td></td></tr>
<tr><td>terminals (general, balanced)</td><td></td></tr>
<tr><td>terminals (general, unbalanced)</td><td></td></tr>
<tr><td>test point</td><td></td></tr>
</table>

ELECTRONIC SYMBOLS *(cont'd)*

thermocouple	
thyristor (diac)	
thyristor (triac)	
transformer (air-core)	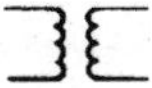
transformer (air-core, adjustable)	
transformer (iron-core)	
transformer (iron-core, adjustable)	
transformer (powdered iron-core)	
transformer (tapped-primary)	
transformer (tapped-secondary)	
transistor (bipolar, npn)	
transistor (bipolar, pnp)	

ELECTRONIC SYMBOLS *(cont'd)*

transistor (junction field-effector, JFET)	
transistor (field-effect, n-channel)	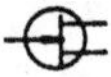
transistor (field-effect, P-channel)	
transistor (metal-oxide, dual-gate)	 
transistor (metal-oxide, single-gate)	 
transistor (photosensitive)	 
transistor (unijunction)	 
tube (diode)	
tube (pentode)	
tube (photomultiplier)	
tube (tetrode)	
tube (triode)	

ELECTRONIC SYMBOLS *(cont'd)*

unspecified unit or component

voltmeter

wattmeter

waveguide (circular)

waveguide (rectangular)

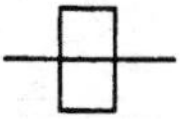

waveguide (flexible)

waveguide (twisted)

ELECTRICAL PREFIXES

Prefixes

Prefixes are used to avoid long expressions of units that are smaller and larger than the base unit. See Common Prefixes. For example, sentences 1 and 2 do not use prefixes. Sentences 3 and 4 use prefixes.

1. A solid-state device draws 0.000001 amperes (A).
2. A generator produces 100,000 watts (W).
3. A solid-state device draws 1 microampere (uA).
4. A generator produces 100 kilowatts (kW).

Converting Units

To convert between different units, the decimal point is moved to the left or right, depending on the unit. See Conversion Table. For example, an electronic circuit has a current flow of .000001 A. The current value is converted to simplest terms by moving the decimal point six places to the right to obtain 1.0µA (from Conversion Table).

.000001. A = 1.0 uA

Move decimal point
6 places to right

Common Electrical Quantities

Abbreviations are used to simplify the expression of common electrical quantities.
See Common Electrical Quantities. For example, milliwatt is abbreviated mW, kilovolt is abbreviated kV, and ampere is abbreviated A.

COMMON PREFIXES

Symbol	Prefix	Equivalent
G	giga	1,000,000,000
M	mega	1,000,000
k	kilo	1000
base unit	—	1
m	milli	.001
u	micro	.000001
n	nano	.000000001

COMMON ELECTRICAL QUANTITIES

Variable	Name	Unit of Measure and Abbreviation
E	voltage	volt - V
I	current	ampere - A
R	resistance	ohm - Ω
P	power	watt - W
P	power (apparent)	volt-amp - VA
C	capacitance	farad - F
L	inductance	henry - H
Z	impedance	ohm - Ω
G	conductance	siemens - S
f	frequency	hertz - Hz
T	period	second - s

CONVERSION TABLE

Initial Units	Final Units						
	giga	mega	kilo	base unit	milli	micro	nano
giga	—	3R	6R	9R	12R	15R	18R
mega	3L	—	3R	6R	9R	12R	15R
kilo	6L	3L	—	3R	6R	9R	12R
base unit	9L	6L	3L	—	3R	6R	9R
milli	12L	9L	6L	3L	—	3R	6R
micro	15L	12L	9L	6L	3L	—	3R
nano	18L	15L	12L	9L	6L	3L	—

ELECTRICAL ABBREVIATIONS

Abbrev.	Term	Abbrev.	Term
A	Amps; armature; anode; ammeter	K	Kilo; cathode
Ag	Silver	L	Line; load
ALM	Alarm	LB-FT	Pounds per feet
AM	Ammeter	LB-IN	Pounds per inch
ARM	Armature	LRC	Locked rotor current
Au	Gold	M	Motor; motor starter contacts
BK	Black	MED	Medium
BL	Blue	N	Nirth
BR	Brown	NC	Normally closed
C	Celsius; centigrade	NO	Normally opened
CAP	Capacitor	NTDF	Nontime-delay fuse
CB	Circuit breaker	O	Orange
CCW	Counterclockwise	OCPD	Overcurrent protection device
CONT	Continuous	OL	Overloads
CPS	Cycles per second	OZ/IN	Ounces per inch
CR	Control relay	P	Power consumed
CT	Current transformer	PSI	Pounds per square inch
CW	Clockwise	PUT	Pull-up torque
D	Diameter	R	Resistance; radius; red; reverse
DP	Double-pole	REV	Reverse
DPDT	Double-pole, double-throw	RPM	Revolutions per minute
EMF	Electromotive force	S	Switch; series; slow; south
F	Fahrenheit; forward; fast	SCR	Silicon controlled rectifier
F	Field; forward	SF	Service factor
FLC	Full-load current	SP	Single-pole
FLT	Full-load torque	SPDT	Single-pole; double-throw
FREQ	Frequency	SPST	Single-pole; single-throw
FS	Float switch	SW	Switch
FTS	Foot switch	T	Terminal; torque
FWD	Forward	TD	Time delay
G	Green; gate	TDF	Time-delay fuse
GEN	Generator	TEMP	Temperature
GY	Gray	V	Volts; violet
H	Transformer, primary side	VA	Voltamps
HP	Horsepower	VAC	Volts alternating current
I	Current	VDC	Volts direct current
IC	Intergrated circuit	W	White; watt
INT	Intermediate; interrupt	W/	With
ITB	Inverse time breaker	X	Transformer secondary side
ITCB	Instantaneous trip circuit breaker	Y	Yellow

CONVERSION FACTORS

Multiply	By	To Obtain
Acres	43,560	Square feet
Acres	1.562×10^{-3}	Square miles
Acre-Feet	43,560	Cubic feet
Amperes per sq cm	6.452	Amperes per sq in.
Amperes per sq in	0.1550	Amperes per sq cm
Ampere-Turns	1.257	Gilberts
Ampere-Turns per cm	2.540	Ampere-turns per in.
Ampere-Turns per in	0.3937	Ampere-turns per cm
Atmospheres	76.0	Cm of mercury
Atmospheres	29.92	Inches of mercury
Atmospheres	33.90	Feet of water
Atmospheres	14.70	Pounds per sq in.
British termal units	252.0	Calories
British thermal units	778.2	Foot-pounds
British thermal units	3.960×10^{-4}	Horsepower-hours
British termal units	0.2520	Kilogram-calories
British thermal units	107.6	Kilogram-meters
British termal units	2.931×10^{-4}	Kilowatt-hours
British thermal units	1,055	Watt-seconds
B.t.u. per hour	2.931×10^{-4}	Kilowatts
B.t.u. per minute	2.359×10^{-2}	Horsepower
B.t.u. per minute	1.759×10^{-2}	Kilowatts
Bushels	1.244	Cubic feet
B.t.u. per hour	2.931×10^{-4}	Kilowatts
B.t.u. per minute	2.359×10^{-2}	Horsepower
B.t.u. per minute	1.759×10^{-2}	Kilowatts
Bushels	1.244	Cubic feet
Centimeters	0.3937	Inches
Circular mils	5.067×10^{-6}	Square centimeters
Circular mils	0.7854×10^{-6}	Square inches
Circular mils	0.7854	Square mils
Cords	128	Cubic feet
Cubic centimeters	6.102×10^{-6}	Cubic inches
Cubic feet	0.02832	Cubic meters
Cubic feet	7.481	Gallons
Cubic feet	28.32	Liters
Cubic inches	16.39	Cubic centimeters
Cubic meters	35.31	Cubic feet
Cubic meters	1.308	Cubic yards
Cubic yards	0.7646	Cubic meters

CONVERSION FACTORS *(cont'd)*

Multiply	By	To Obtain
Degrees (angle)	0.01745	Radians
Dynes	2.248×10^{-6}	Pounds
Ergs	1	Dyne-centimeters
Ergs	7.37×10^{-6}	Foot-pounds
Ergs	10^{-7}	Joules
Farads	10^{6}	Microfarads
Fathoms	6	Feet
Feet	30.48	Centimeters
Feet of water	.08826	Inches of mercury
Feet of water	304.8	Kg per square meter
Feet of water	62.43	Pounds per square ft.
Feet of water	0.4335	Pounds per square in.
Foot-pounds	1.285×10^{-2}	British thermal units
Foot-pounds	5.050×10^{-7}	Horsepower-hours
Foot-pounds	1.356	Joules
Foot-pounds	0.1383	Kilogram-meters
Foot-pounds	3.766×10^{-7}	Kilowatt-hours
Gallons	0.1337	Cubic feet
Gallons	231	Cubic inches
Gallons	3.785×10^{-3}	Cubic meters
Gallons	3.785	Liters
Gallons per minute	2.228×10^{-3}	Cubic feet per sec.
Gausses	6.452	Lines per square in.
Gilberts	0.7958	Ampere-turns
Henries	10^{3}	Millihenries
Horsepower	42.41	Btu per min
Horsepower	2,544	Btu per hour
Horsepower	550	Foot-pounds per sec
Horsepower	33,000	Foot-pounds per min
Horsepower	1.014	Horsepower (metric)
Horsepower	10.70	Kg calories per min
Horsepower	0.7457	Kilowatts
Horsepower (boiler)	33,520	Btu per hour
Horsepower-hours	2,544	British thermal units
Horsepower-hours	1.98×10^{6}	Foot-pounds
Horsepower-hours	2.737×10^{5}	Kilogram-meters
Horsepower-hours	0.7457	Kilowatt-hours

CONVERSION FACTORS *(cont'd)*

Multiply	By	To Obtain
Inches	2.540	Centimeters
Inches of mercury	1.133	Feet of water
Inches of mercury	70.73	Pounds per square ft.
Inches of mercury	0.4912	Pounds per square in.
Inches of water	25.40	Kg per square meter
Inches of water	0.5781	Ounces per square in.
Inches of water	5.204	Pounds per square ft
Joules	9.478×10^{-4}	British thermal units
Joules	0.2388	Calories
Joules	10^{7}	Ergs
Joules	0.7376	Foot-pounds
Joules	2.778×10^{-7}	Kilowatt-hours
Joules	0.1020	Kilogram-meters
Joules	1	Watt-seconds
Kilograms	2.205	Pounds
Kilogram-calories	3.968	British thermal units
Kilogram meters	7.233	Foot-pounds
Kg per square meter	3.281×10^{-3}	Feet of water
Kg per square meter	0.2048	Pounds per square ft
Kg per square meter	1.422×10^{-3}	Pounds per square in.
Kilolines	10^{3}	Maxwells
Kilometers	3.281	Feet
Kilometers	0.6214	Miles
Kilowatts	56.87	Btu per min
Kilowatts	737.6	Foot-pounds per sec
Kilowatts	1.341	Horsepower
Kilowatts-hours	3409.5	British thermal units
Kilowatts-hours	2.655×10^{6}	Foot-pounds
Knots	1.152	Miles
Liters	0.03531	Cubic feet
Liters	61.02	Cubic inches
Liters	0.2642	Gallons
Log N_e or in N	0.4343	Log_{10} N
Log N	2.303	Log_e N or in N
Lumens per square ft	1	Footcandles
Maxwells	10^{-3}	Kilolines
Megalines	10^{6}	Maxwells

CONVERSION FACTORS *(cont'd)*

Multiply	By	To Obtain
Megaohms	10^6	Ohms
Meters	3.281	Feet
Meters	39.37	Inches
Meter-kilograms	7.233	Pound-feet
Microfarads	10^{-6}	Farads
Microhms	10^{-6}	Ohms
Microhms per cm cube	0.3937	Microhms per in. cube
Microhms per cm cube	6.015	Ohms per mil foot
Miles	5,280	Feet
Miles	1.609	Kilometers
Miner's inches	1.5	Cubic feet per min
Ohms	10^{-6}	Megohms
Ohms	10^6	Microhms
Ohms per mil foot	0.1662	Microhms per cm cube
Ohms per mil foot	0.06524	Microhms per in. cube
Poundals	0.03108	Pounds
Pounds	32.17	Poundals
Pound-feet	0.1383	Meter-Kilograms
Pounds of water	0.01602	Cubic feet
Pounds of water	0.1198	Gallons
Pounds per cubic foot	16.02	Kg per cubic meter
Pounds per cubic foot	5.787×10^{-4}	Pounds per cubic in.
Pounds per cubic inch	27.68	Grams per cubic cm
Pounds per cubic inch	2.768×10^{-4}	Kg per cubic meter
Pounds per cubic inch	1.728	Pounds per cubic ft
Pounds per square foot	0.01602	Feet of water
Pounds per square foot	4.882	Kg per square meter
Pounds per square foot	6.944×10^{-3}	Pounds per sq. in.
Pounds per square inch	2.307	Feet of water
Pounds per square inch	2.036	Inches of mercury
Pounds per square inch	703.1	Kg per square meter
Radians	57.30	Degrees
Square centimeters	1.973×10^5	Circular mils
Square Feet	2.296×10^{-5}	Acres
Square Feet	0.09290	Square meters
Square inches	1.273×10^6	Circular mils
Square inches	6.452	Square centimeters
Square Kilometers	0.3861	Square miles

CONVERSION FACTORS *(cont'd)*

Multiply	By	To Obtain
Square meters	10.76	Square feet
Square miles	640	Acres
Square miles	2.590	Square kilometers
Square Milimeters	1.973×10^3	Circular mils
Square mils	1.273	Circular mils
Tons (long)	2,240	Pounds
Tons (metric)	2,205	Pounds
Tons (short)	2,000	Pounds
Watts	0.05686	Btu per minute
Watts	10^7	Ergs per sec
Watts	44.26	Foot-pounds per min.
Watts	1.341×10^{-3}	Horsepower
Watts	14.34	Calories per min.
Watts-hours	3.412	British thermal units
Watts-hours	2,655	Footpounds
Watts-hours	1.341×10^{-3}	Horsepower-hours
Watts-hours	0.8605	Kilogram-calories
Watts-hours	376.1	Kilogram-meters
Webers	10^8	Maxwells

CHAPTER 2
NETWORKING

LAN TOPOLOGIES

- The geometric, electrical or physical arrangement describing a communications network.
- A sketch of how the network terminals are arranged.
- The most common are: bus, ring and star.

BUS

- Bus Topology
  - Terminals are tapped into a single transmission line; paralleled.
  - Network is controlled by a terminal addressing scheme.
- Advantages
  - Low cable usage
  - Simple cable arrangement
  - Terminals can be actively installed
  - Terminal failure will not block network traffic
- Disadvantages
  - Design limitations
  - System expansion limitations
  - Impedance problems

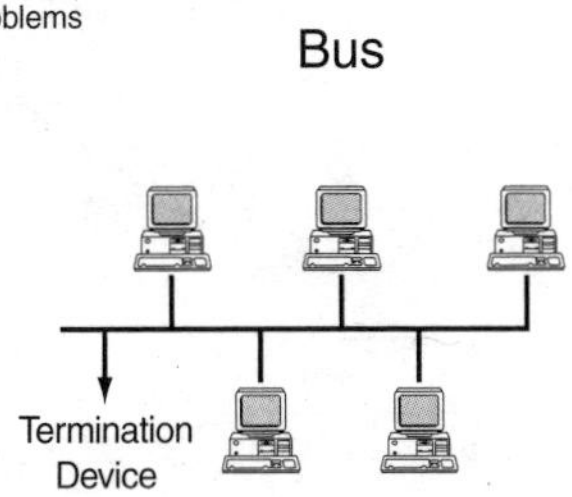

RING

- Ring Topology
  - Terminals are wired in a ring configuration; series.
- Advantages
  - Minimal cable usage.
  - Terminals can be added without a new cable run.
  - Simple wiring scheme.
- Disadvantages
  - A break in the ring will cause the entire system to fail.
  - Difficult to troubleshoot.
  - Wiring changes can affect the entire network.

Ring

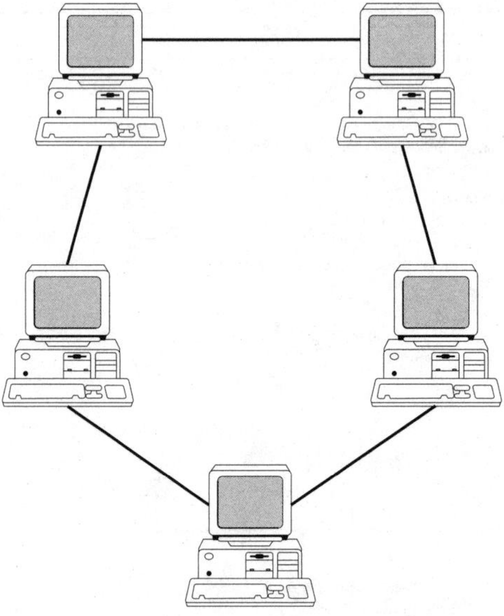

STAR

- Star Topology
  - Terminals are wired to one centralized location
- Advantages
  - Ease of troubleshooting
  - Simple wiring management
  - Minimal preplanning
- Disadvantages
  - High cable usage
  - Distance limitations

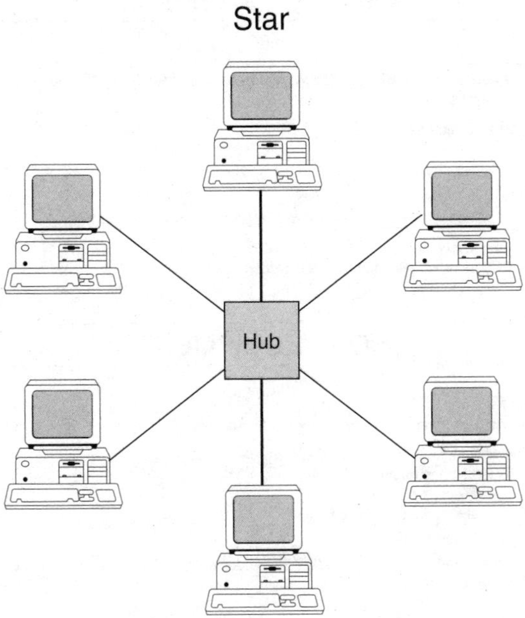

LAN PROTOCOLS

- Definition
  - A set of rules that define how network terminals can communicate across a given network.
- OSI Model
  - OSI (open systems interconnect)
  - 7 Layers (application, presentation, session, transport, network, data link, physical)
- Examples:
  - Ethernet (IEEE 802.3)
  - Token Ring (IEEE 802.5)
  - FDDI (ANSI X379.5)

TERMINAL EQUIPMENT

- Definition:
  - a network terminal is a device that can send and/or receive information. Sometimes also referred to as a "node".
- Terminal Hardware
  - physical device
  - includes adapter cards and mounting cables
- Terminal Software
  - must be loaded into the device to make it compatible with the network
- Examples:
  - PCs, printers, telephones, video cameras, etc.

HOST EQUIPMENT

- Definition:
  - equipment that is commonly shared by many or all terminals on the network.
- Installation
  - located in the equipment room
  - usually very expensive
  - correct installation is critical
- Examples
  - mainframe computers, multiplexers, modems, PBXs, key systems, voice mail systems

TYPICAL NETWORK STRUCTURE

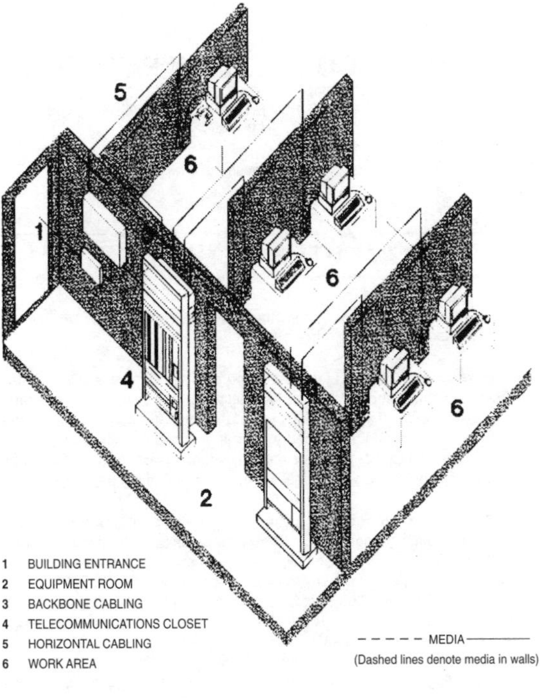

1 BUILDING ENTRANCE
2 EQUIPMENT ROOM
3 BACKBONE CABLING
4 TELECOMMUNICATIONS CLOSET
5 HORIZONTAL CABLING
6 WORK AREA

– – – – – MEDIA ————
(Dashed lines denote media in walls)

NETWORK COMMUNICATION MEANS

1. Twisted pair cables.
2. Coaxial cables.
3. Optical fiber cables.
4. Radio waves.
5. Infrared light.
6. Electronic signals sent through power lines.

NETWORK CONNECTION PATTERNS

1. A star pattern.
2. A ring pattern.
3. A bus pattern.
4. A mesh pattern.

COMPUTER NETWORKS

There are two basic types of computer networking models: *Centralized* computing and *Client/Server* computing:

Centralized Computing: In the past, corporate data communications involved accessing a central computer. Everybody went to this one computer to take care of a particular task or business process. Input to the computer was made using interactive (dumb) terminals. Later, smart terminals provided for batched input to the mainframe. These types of terminals are often found in retail chains where stores download sales information to the mainframe at the end of the day.

Client/Server Computing: The general availability of microprocessor-based Personal Computers changed the way networking was done. With more intelligent terminals, most, or in some cases all, of the processing load is performed at the desk through a Personal Computer, and not at the mainframe.

Along with the Client/Server computer model came new methods of getting computers to talk to one another. A high-speed transmission media was needed, called a Local Area Network (LAN). Also, computers had to talk the same network language, to form a Network Operating System (NOS). With a NOS, your computer's operating system is integrated into the network. These systems have been developed and made widely available by computers such as DEC, IBM, Novell, LanTastic, 3Com, Xerox, Banyan-Vines, and Microsoft.

CENTRALIZED COMPUTING MODEL

Relies heavily on WAN (wide area network) technologies

Well-suited for mission—critical information

High computer cost

Low end-user equipment costs

Lower network management costs

Higher transmission facility costs

Lacks flexibility and customization

CLIENT/SERVER COMPUTING MODEL

Relies on both LAN and WAN technologies

Flexible deployment—easily customized

Low computer cost

Increased end-user equipment costs

Lower transmission facility costs

Increased network management costs

COMMON ARCHITECTURES

- **Small, single segment networks.** Usually 100 or fewer users. The common topologies are Ethernet and Token-Ring. The network usually has no more than two or three servers.
- **Medium-sized, collapsed backbone networks.** Usually 1,000 or fewer users. Common topologies are Ethernet and Token-Ring. A single router, or just a couple of routers, sit as the backbone of the network and provide connectivity for the network. The network usually has about ten servers.
- **Large, high-speed connected networks.** Usually more than 1,000 users and often involving more than one building or a number of floors of a large office building. Desktop connectivity is still with Ethernet and Token-Ring, backbone is most often FDDI. Routers sit on the high-speed backbone and provide connectivity to attached Ethernet and Token-Ring segments.

NETWORK DEVICES MOST COMMONLY USED IN ETHERNETS

- A **repeater** receives and then immediately retransmits each bit. It has no memory and does not depend on any particular protocol. It duplicates everything, including the collisions.

- A **bridge** receives the entire message into memory. If the message was damaged by a collision or noise, it is discarded. If the bridge knows that the message was being sent between two stations on the same cable, it discards it. Otherwise, the message is queued up and will be retransmitted on another Ethernet cable. The bridge has no address. Its actions are transparent to the client and server workstations.

- A **router** acts as an agent to receive and forward messages. The router has an address and is known to the client or server machines. Typically, machines directly send messages to each other when they are on the same cable, and they send the router messages addressed to another zone, department, or subnetwork. Routing is a function specific to each protocol. For example, on an IPX system, the Novell server can act as a router. For SNA, an APPN Network Node does the routing. TCP/IP can be routed by dedicated devices, UNIX workstations, or OS/2 servers.

ETHERNET FRAME TYPES

Ethernet 802.3 (Raw)

This is the original (and default) frame type used by NetWare. IT CAN ONLY SUPPORT NOVELL IPX/SPX TRAFFIC! The frame is similar to that described in 802.3 except that it does not contain the Logical Link Control (LLC) information in the packet.

Preamble: 8 bytes
Destination Address: 6 bytes
Source Address: 6 bytes
Length Field: 2 bytes
Data Field: Between 46 and 1500 bytes
Pad Characters: Variable, stuffs data field up to 46 bytes
Frame Check Sequence: 4 bytes
Min Frame Length: 64 bytes
Max Frame Length: 1518 bytes (not including Preamble)

Ethernet 802.2

This frame includes fields from 802.3 and 802.2 (Logical Link Control) and can support the Novell IPX/SPX and FTAM (File Transfer, Access, and Management) protocols. The frame parameters are identical to those listed above, EXCEPT that the first three bytes of the data field are used to indicate 802.2 header (LLC) information.

Preamble: 8 bytes
Destination Address: 6 bytes
Source Address: 6 bytes
Length Field: 2 bytes
Data Field: Between 46 and 1500 bytes (including LLC)
Pad Characters: Variable, stuffs data field up to 46 bytes
Frame Check Sequence: 4 bytes
Min Frame Length: 64 bytes
Max Frame Length: 1518 bytes (not including Preamble)

The LLC field consists of:

Destination Service Access Point (DSAP): 1 byte (NetWare 0xE0)
Source Service Access Point (SSAP): 1 byte (NetWare 0xE0)
Control Field: 1 byte (NetWare 0x03)

NetWare IPX/SPX packets will assign a hexadecimal value of E0 to the DSAP and SSAP fields and a hexadecimal value of 03 to the Control field. The “03” Control value indicates an unnumbered 802.2 layer.

ETHERNET FRAME TYPES *(cont'd)*

Ethernet II

Again, a similar frame type, EXCEPT that the two-byte Length field has been replaced with a two-byte Type field (Ethertype). Ethernet II frames can support Novell IPX/SPX, TCP/IP, and AppleTalk Phase 1 protocols. Ethernet II frames do not use a LLC header in the data field.

Preamble: 8 bytes
Destination Address: 6 bytes
Source Address: 6 bytes
Ethernet Type: 2 bytes (Novell 0x81-37)
Data Field: Between 46 and 1500 bytes
Pad Characters: Variable, stuffs data field up to 46 bytes
Frame Check Sequence: 4 bytes
Min Frame Length: 64 bytes
Max Frame Length: 1518 bytes (not including Preamble)

The Ethernet Type field is coded with hexadecimal 8137 for transport of NetWare IPX/SPX packets.

Ethernet SNAP

Sub-Network Access Protocol (SNAP) is similar to 802.2, with LLC parameters, but with expanded LLC capabilities. Ethernet SNAP can support IPX/SPX, TCP/IP, and AppleTalk Phase 2 protocols.

Preamble: 8 bytes
Destination Address: 6 bytes
Source Address: 6 bytes
Length Field: 2 bytes
Data Field: Between 46 and 1500 bytes (including LLC)
Pad Characters: Variable, stuffs data field up to 46 bytes
Frame Check Sequence: 4 bytes
Min Frame Length: 64 bytes
Max Frame Length: 1518 bytes (not including Preamble)

ETHERNET FRAME TYPES *(cont'd)*

The LLC field (the first eight bytes in the data field) consists of:

Destination Service Access Point (DSAP): 1 byte (0xAA)
Source Service Access Point (SSAP): 1 byte (0xAA)
Control Field: 1 byte (NetWare 0x03)
Organizational Code: 3 bytes (0x00-00-00)
Ethernet Type: 2 bytes (NetWare 0x81-37)

Hexadecimal AA (decimal 170) values are usually employed for the DSAP and SSAP values. Netware uses a hexadecimal 03 in the SNAP Control field and will use the 0x81-37 Ethernet Type value. Usually, the Organizational Code is coded as all 0s (0x00-00-00) and NetWare is no exception.

DISTINGUISHING BETWEEN ETHERNET FRAME TYPES

1) Receive good frame.
2) Analyze frame. Perform the following steps, in order:

If the EtherType/Length value is greater than 0x05-DC (decimal 1500), then process the frame as Ethernet II. Any EtherType value greater than 0x05-DC (such as 0x0800 for IP or 0x81-37 for NetWare IPX/SPX) will be interpreted as an Ethernet II frame.

If the IPX header (0xFF-FF) follows the Length field, the frame is interpreted as a 802.3 (Raw) frame with Netware IPX/SPX traffic. Standard SSAP and DSAP values do not include hexadecimal FF, so the 802.3 (Raw) frame can be distinguished from LLC frames (Ethernet SNAP, 802.2).

Next, the byte following the length field (DSAP) is examined. If the value is 0xAA, the frame is interpreted as a SNAP frame. Otherwise, it is interpreted as a 802.2 frame.

GIGABIT ETHERNET PACKETS

PREAMBLE	SFD	DA	SA	LENGTH OF DATA FIELD	DATA	FCS
7 Byte	1 Byte	6 Byte	6 Byte	2 Byte	46-1500 Byte	4 Byte

Frame (6+6+2+46+4 = 64 bytes min size)
(6+6+2+1500+4 = 1518 bytes max size)

Ethernet Packet

BIT ERROR RATE

$$\textbf{BIT ERROR RATE} = \frac{\text{TRANSMITTED FRAMES – RECEIVED FRAMES}}{\text{TRANSMITTED FRAMES (10 Billion)}}$$

OR

$$\textbf{BIT ERROR RATE} = \frac{\text{TOTAL NUMBER OF ERRORS (Packer Loss, CRC, Etc.)}}{\text{TRANSMITTED FRAMES (10 Billion)}}$$

$$\underset{\text{(IN BYTES)}}{\textbf{BIT ERROR RATE}} = \frac{\text{TOTAL NUMBER OF ERRORS x 8}}{\text{TRANSMITTED FRAMES x 8}}$$

ETHERNET LAN CONFIGURATIONS

10Base5

This interface is also known as "Thick Ethernet" and is based upon the use of thick, inflexible, coaxial cable. The coaxial cable has an impedance of 50 ohms, and is terminated with Male "Type N" coax connectors. The center conductor is solid.

- Taps per segment: 100
- Maximum separation: 5 segments, 4 repeaters
- Maximum segment length: 500 meters
- Total network span: 2500 meters
- Minimum length between transceivers: 2.5 meters
- Max transceiver drop length: 50 meters

Each end of a coaxial segment must be terminated into a 50 ohm impedance.

10Base2

This interface is also known as "Thin Ethernet" and is based upon the use of thin, flexible, coaxial cable. The coaxial cable has an impedance of 50 ohms, and is terminated with Male "BNC" (Bayonet-Niell and Concelman) type coax connectors. The center conductor is stranded.

- Taps per segment: 30
- Maximum separation: 5 segments, 4 repeaters
- Maximum segment length: 185 meters
- Total network span: 925 meters
- Minimum length between transceivers: 0.5 meters

Each end of a coaxial segment must be terminated into a 50 ohm impedance. There can be no more than TWO 50 ohm terminators per segment, or else the collision detection function may not operate properly. Multiport repeaters may include a 50 ohm termination.

10Base T

The Physical Interconnection—The following diagram depicts the physical interconnection of a PC (Workstation) to the Ethernet LAN:

Functional Overview: Twisted pair cable is used as the transport media. This is known as a Type "10Base-T" Ethernet. In these configurations, each station (PC) is connected to a central "repeater" or "hub" (Star Network). The cable length may be up to 100 meters when Category 3 cable is used.

Items C and D (MAU and AUI) are associated with a "Transceiver", if applicable. Usually, these functions are onboard the Ethernet board in the PC and no Transceiver is required. Typical pinouts for the RJ-45 jack are:

1 = DET Transmit Data (+)
2 = DTE Transmit Data (−)
3 = DTE Receive Data (+)
6 = DTE Receive Data (−)

Typical Transceiver pinouts are:

1 = Collision Shield
2 = Collision (+)
3 = Transmit (+)
4 = Receive Shield
5 = Receive (+)
6 = Power Return
7 = Not Used
8 = Not Used
9 = Collision (−)
10 = Transmit (−)
11 = Transmit Shield
12 = Receive (−)
13 = +12 VDC
14 = Voltage Shield
15 = Not Used

Item E, the MAC (Media Access Control) function resides onboard the Ethernet card. MAC operation controls the electrical functions associated with Carrier Sense Multiple Access (CSMA)/Collision Detection (CD) and the output/decoding of Manchester-coded 10 MBPS Ethernet HDLC frames.

DEFINITIONS OF NETWORK COMPONENTS

Segment—The smallest piece of a network on which stations can exchange data without intervention from another intelligent device.

Extended Segment—A number of segments joined together by bridges. Any broadcast or multicast made on the extended segment should be seen by all stations on an extended segment.

Network—The term itself has come to be rather ambiguous, referring to a segment, extended segment or internetwork. We often call any of these "The Network."

Internetwork—A set of segments or extended segments joined together by a router.

Unicast Packet—A data packet addressed to a single station. An example might be data from a client to a server.

Multicast Packet—A data packet addressed to a group of stations. The destination address is formed in such a way that stations realize that the packet may be destined for many other stations.

Broadcast Package—A data packet addressed to any and all stations on the local segment. Broadcasts are often used by stations who have just joined the network—broadcasts are made to find out information about the segment that has just been joined.

Repeater—A device that facilitates connecting stations onto the segment. It does not understand network addresses—it merely copies data bit by bit from and to the physical media to which it is attached. On Token-Ring segments, this device is often called a Media Access Unit or MAU. A repeater is not considered an intelligent device.

Bridge—A bridge is used to connect two or more similar segments together (for example, Token-Ring to Token-Ring or Ethernet to Ethernet). A bridge has two purposes. The first is to extend the length and number of stations that a segment can support. Secondly, the bridge reduces overall traffic flow by only passing data packets that are not destined for a hardware address on a local segment. All broadcast and multicast traffic must cross a bridge—since no true destination can be known. In recent years, bridging technology has been used between dissimilar media (for example, Ethernet to FDDI), this sometimes may cause problems as we will see later. A bridge is considered an intelligent device. (See also bridging.)

Router—Sometimes called a gateway, it is used to connect two or more (potentially extended) segments. The segments may be similar or dissimilar. Routing information beyond the hardware address must be contained within the data packet. Virtually no broadcasts or multicasts are ever propagated across a router since no exact destination information is typically contained within these packets. Hardware addresses have only local significance to a router—higher level routing information is globally significant. (See also routing.)

NETWORK TERMS

Baseband network. A baseband network is one that provides a single channel for communications across the physical medium (cable), so that only one device can transmit at a time. Devices on a baseband network, such as Ethernet, are permitted to use all the available bandwidth for transmission, and the signals they transmit do not need to be multiplexed onto a carrier frequency. An analogy is a single phone line: Only one person can talk at a time—if more than one person wants to talk everyone has to take turns.

Broadband network. A baseband network is in many ways the opposite of a baseband network. With broadband, the physical cabling is virtually divided into several different channels, each with its own unique carrier frequency, using a technique called *frequency division modulation.* These different frequencies are multiplexed onto the network cabling in such a way to allow multiple simultaneous "conversations" to take place. The effect is similar to having several virtual networks traversing a single piece of wire. Network devices tuned to one frequency can't hear the signal on other frequencies, and visa-versa. Cable-TV is the best example of a broadband network, with multiple conversations (channels) transmitted simultaneously over a single cable; you pick which one you want to see by selecting the frequencies.

OSI Model. The *Open Systems Interconnect* (OSI) reference model is the ISO (*International Standards Organization*) structure for network architecture. This Model outlines seven areas, or layers, for the network. These layers are (from highest to lowest:

7. *Applications:* Where the user applications software lies. Such issues as file access and transfer, virtual terminal emulation, interprocess communication and the like are handled here.

6. *Presentation:* Differences in data representation are dealt with at this level. For example, UNIX-style line endings (CR only) might be converted to MS-DOS style (CRLF), or EBCIDIC to ASCII character sets.

5. *Session:* Communications between applications across a network is controlled at the session layer. Testing for out-of-sequence packets and handling two-way communication are handled here.

4. *Transport:* Makes sure the lower three layers are doing their job correctly, and provides a transparent, logical data stream between the end user and the network service. This is the lower layer that provides local user services.

3. *Network:* This layer makes certain that a packet sent from one device to another actually gets there in a reasonable period of time. Routing and flow control are performed here. This is the lowest layer of the OSI model that can remain ignorant of the physical network.

2. *Data Link:* This layer deals with getting data packets on and off the wire, error detection and correction, and retransmission. This layer is generally broken into two sub-layers: The LLC (Logical Link Control) on the upper half, which does the error checking, and the MAC (Medium Access Control) on the lower half, which deals with getting the data on and off the wire.

1. *Physical:* The nuts and bolts layer. Here is where the cable, connector and signaling specifications are defined.

10Base5, 10BaseT, 10Base 2, 10Broad36, etc. These are the IEEE names for the different physical types of Ethernet. The "10" stands for signaling speed: 10MHz. "Base" means Baseband, "broad" means broadband. Initially, the last section as intended to indicate the maximum length of an unrepeated cable segment in hundreds of meters. This convention was modified with the introduction of 10BaseT, where the T means twisted pair, and 10BaseF where the F means fiber.

- 10Base2 Is 10MHz Ethernet running over thin, 50 Ohm baseband coaxial cable. 10Base2 is also commonly referred to as thin-Ethernet.
- 10Base5 Is 10MHz Ethernet running over standard (thick) 50 Ohm baseband coaxial cabling.
- 10BaseF Is 10MHz Ethernet running over fiber-optic cabling.
- 10BaseT Is 10MHz Ethernet running over unshielded, twisted-pair cabling.
- 10Broad36 Is 10MHz Ethernet running through a broadband cable.

CHAPTER 3
STRUCTURED CABLING

HISTORY OF STRUCTURED CABLING

As computer networks originated every manufacturer's system was different—you ran cable one way for an IBM system, a different way for a DEC system, and a third way for an AT&T system. Not only did cable routes vary, but cable types varied. Eventually, the manufacturers got together (through the EIA/TIA organization), and came up with a generic cable routing pattern that they would all use called EIA/TIA 586, and gave it the name "Structured Cabling." 568 has been revised several times, but the basic structure remains the same. *Structured cabling* refers to a network cabling system that is designed and installed according to the pre-set standards of EIA/TIA 568. The benefits of structured cabling are the following:

- Buildings, new or re-furbished, are pre-wired without needing to know future occupant's data communication needs.
- Future growth and re-configuration accommodated by pre-defined topologies and physical specifications, such as distances.
- Support of multi-vendor products, including cables, connectors, jacks, plugs, adapters, baluns, and patch-panels.
- Voice, video and all other data transmissions are integrated.
- Cable plant easily managed and faults readily isolated.
- Cabling work can be done while other building work is taking place.

Parameters of EIA/TIA 568

- Up to 50,000 users
- Facilities up to 10 million square feet
- 90 meter horizontal distance limit between closet and desktop
- 4 pairs of conductors to each outlet—all must be terminated
- 25-pair cables may *not* be used (crosstalk problems)
- May not use old wiring already in place
- Bridge taps and standard telephone wiring schemes may not be used
- Requires careful installation procedures
- Requires extensive testing procedures

TYPICAL 568 CABLING LAYOUT

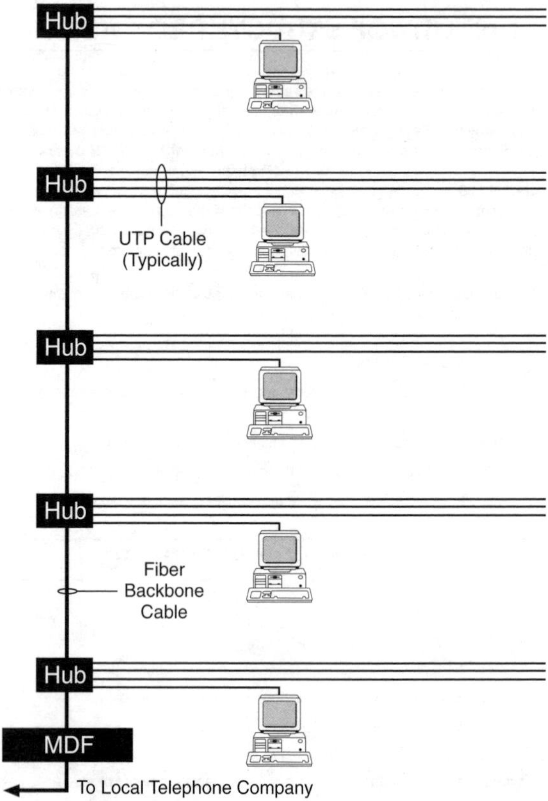

TYPICAL 568 CABLING LAYOUT *(cont'd)*

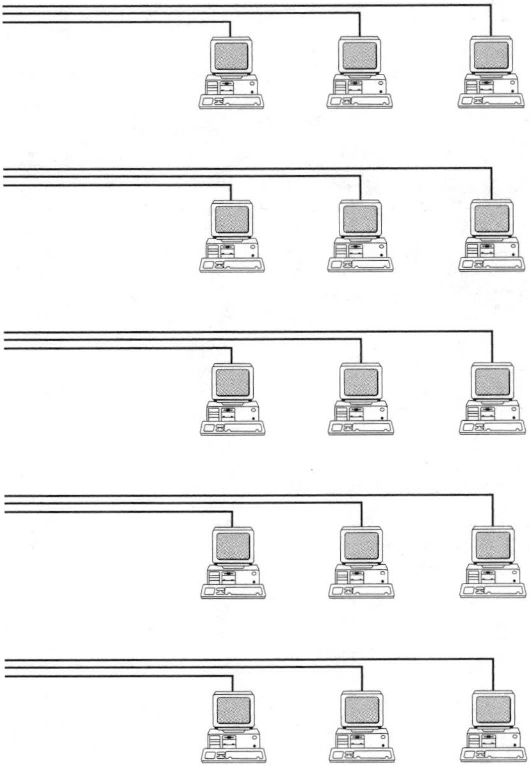

DISTANCE LIMITS FOR HORIZONTAL CABLING

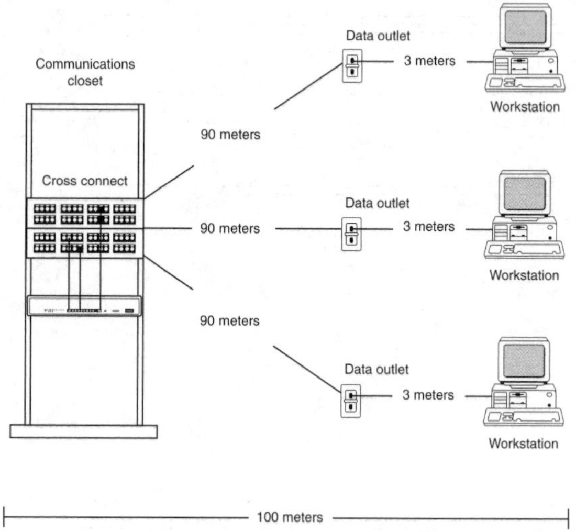

In addition to 90 meters of horizontal cable, 10 meters is allowed for work area and communications closet cables.

EIA/TIA 568-B HORIZONTAL CABLES

	Cat. 3	Cat. 5 Annex N		Cat. 5e	Cat. 6 Draft 11
	Characteristic Impedence, Ohms, using a smoothing functions over the bandwidth				
1 - 16 MHz	100 ± 15%				
1 - 100 MHz		100 ± 15%		100 ± 15%	
1 - 250 Mhz					100 ± 15%
	Structural Return Loss (SRL), dB and Return Loss (RL), dB				
	SRL (Min)	SRL (Min)	RL (Min)	RL (Min)	RL (Min)
1 - 10 MHz	12	23	17 + 3Log(f)	20 + 5log(f)	20 + 5log(f)
10 - 16 MHz	12	23			
10 -20 MHz	12 - 10log(f/10)		20	25	25
16 - 20 MHz		23			
20 - 100 MHz		16 - 10log(f/100)	20 - 7log(f/20)	25 - 7log(f/20)	
20 - 250 MHz					25 - 7log(f/20)

EIA/TIA 568-B HORIZONTAL CABLES *(cont'd)*

	Cat. 3	Cat. 5 Annex N	Cat. 5e	Cat. 6 Draft 11
Insertion Loss dB/100m				
Limits are defined by formula: Insertion Loss (f) = k1sqrt(f) + k2(f) + k3/sqrt(f) for all frequencies in the range from 0.772 MHz to the highest referenced frequency where the constants k1, k2, & k3 are defined for each category cable.				
k_1	2.320	1.967	1.967	1.808
k_2	0.238	0.023	0.023	0.017
k_3	0.000	0.050	0.050	0.200
0.772 Mhz	2.2	1.8	1.8	1.8
1.0 MHz	2.6	2.0	2.0	2.0
4.0 MHz	5.6	4.1	4.1	3.8
8.0 Mhz	8.5	5.8	5.8	5.3
10.0 MHz	9.7	6.5	6.5	6.0
16.0 Mhz	13.1	8.2	8.2	7.6
20.0 MHz		9.3	9.3	8.5

EIA/TIA 568-B HORIZONTAL CABLES *(cont'd)*

	Cat. 3	Cat. 5 Annex N	Cat. 5e	Cat. 6 Draft 11
25.0 MHz		10.4	10.4	9.5
31.25 MHz		11.7	11.7	10.7
62.5 MHz		17.0	17.0	15.4
100.0 MHz		22.0	22.0	19.8
200.0 MHz				29.0
250.0 MHz				32.8

NEXT

Near-End Crosstalk Loss (NEXT), dB - Limits are defined by the formula:

Cat 3 NEXT (f) = 23.2 - 15log(f/16) for all frequencies in the range from 0.772 - 16 MHz

Cat 5 NEXT(f) = 32.3 - 15 log(f/100) for all frequencies in the range from 0.772 - 100 Mhz

Cat 5e NEXT(f) = 35.3 - 15log(f/100) for all frequencies in the range from 0.772 - 100 Mhz

Cat 6 Next(f) = 44.3 - 15 log(f/100) for all frequencies in the range from 0.772 - 250 MHz

EIA/TIA 568-B HORIZONTAL CABLES *(cont'd)*

	Cat. 3	Cat. 5 Annex N	Cat. 5e	Cat. 6 Draft 11
0.772 Mhz	42.9	64.0	67.0	76.0
1.0 MHz	41.3	62.3	65.3	74.3
4.0 MHz	32.2	53.3	56.3	65.3
8.0 Mhz	27.7	48.8	51.8	60.8
10.0 MHz	26.3	47.3	50.3	59.3
16.0 Mhz	23.2	44.2	47.3	56.2
20.0 MHz		42.8	45.8	54.8
25.0 MHz		41.3	44.3	53.3
31.25 MHz		39.9	42.9	51.9
62.5 MHz		35.4	38.4	47.4
100.0 Mhz		32.3	35.3	44.3
200.0 MHz				39.8
250.0 MHz				38.3

EIA/TIA 568-B HORIZONTAL CABLES *(cont'd)*

	Cat. 3	Cat. 5 Annex N	Cat. 5e	Cat. 6 Draft 11
PSNEXT				
Power Sum Near-End Crosstalk Loss (NEXT), dB - Limits are defined by the formula:				
Cat 5PSNext(f) = 32.3 - 15log(f/100) for all frequencies in the range from 0.772 - 100 MHz				
Cat 5e PSNEXT(f) = 32.3 - 15log(f/100) for all frequencies in the range from 0.772 - 100 MHz				
Cat 6 PSNEXT(f) = 42.3 - 15log(f/100) for all frequencies in the range from 0.772 - 250 MHz				
0.772 MHz		64.0	64.0	74.0
1.0 MHz		62.3	62.3	72.3
4.0 Mhz		53.3	53.3	63.3
8.0 MHz		48.8	48.8	58.8
10.0 MHz		47.3	47.3	57.3
16.0 MHz		44.2	44.2	54.2
20.0 MHz		42.8	42.8	52.8

EIA/TIA 568-B HORIZONTAL CABLES *(cont'd)*

	Cat. 3	Cat. 5 Annex N	Cat. 5e	Cat. 6 Draft 11
25.0 MHz		41.3	41.3	51.3
31.25 MHz		39.9	39.9	49.9
62.5 MHz		35.4	35.4	45.4
100.0 MHz		32.3	32.3	42.3
200.0 MHz				37.8
250.0 MHz				36.3
ELFEXT				
Equal Level Far-End Crosstalk (ELFEXT), dB - Limits are Defined by the formula:				
Cat 5 ELFEXT(f) = 20.8 - 20log(f/100) for all frequencies in the range from 0.772 - 100 MHz				
Cat 5e ELFEXT(f) = 23.8 - 20log(f/100) for all frequencies in the range from 0.772 - 100 MHz				
Cat 6 ELFEXT(f) = 27.8 - 20log(f/100) for all frequencies in the range from 0.772 - 250 MHz				
0.772 MHz		63.0	66.0	70.0

EIA/TIA 568-B HORIZONTAL CABLES *(cont'd)*

	Cat. 3	Cat. 5 Annex N	Cat. 5e	Cat. 6 Draft 11
1.0 MHz		60.8	63.8	67.8
4.0 MHz		48.8	51.7	55.8
8.0 MHz		42.7	45.7	49.7
10.0 MHz		40.8	43.8	47.8
16.0 MHz		36.7	39.7	43.7
20.0 MHz		34.8	37.7	41.8
25.0 MHz		32.8	35.8	39.8
31.25 MHz		30.9	33.9	37.9
62.5 MHz		24.9	27.8	31.9
100.0 MHz		20.8	23.8	27.8
200.0 MHz				21.8
250.0 MHz				19.8

EIA/TIA 568-B HORIZONTAL CABLES *(cont'd)*

	Cat. 3	Cat. 5 Annex N	Cat. 5e	Cat. 6 Draft 11
PSELFEXT				
Power Sum Equal Level Far-End Crosstalk (PSELFEXT), dB - Limits are Defined by the formula:				
Cat 5e PSELFEXT(f) = 20.8 - 20log(f/100) for all frequencies in the range from 1 - 100 MHz				
Cat 6 PSELFEXT(f) = 24.8 - 20log(f/100) for all frequencies in the range from 1 - 250 MHz				
0.772 MHz			63.0	67.0
1.0 MHz			60.8	64.8
4.0 MHz			48.7	52.8
8.0 MHz			42.7	46.7
10.0 MHz			40.8	44.8
16.0 MHz			30.7	40.7
20.0 NHz			34.7	38.8
25.0 MHz			32.8	36.8
31.25 MHz			30.9	34.9
62.5 MHz			24.8	28.9

EIA/TIA 568-B HORIZONTAL CABLES *(cont'd)*

	Cat. 3	Cat. 5 Annex N	Cat. 5e	Cat. 6 Draft 11
100.0 MHz			20.8	24.8
200.0 MHz				18.8
250.0 MHz				16.8
Cable Propagation Delay (Maximum ns/100 m)				
		534 + 36/f	534 + 36/f	534 + 36/f
Cable Propagation Delay Skew (Maximum ns/100m)				
		45	45	45
DC Resistance (Maximum ?/100m)				
	9.38	9.38	9.38	9.38
DC Resistance Unbalance (Maximum %)				
5	5	5	5	5
Mutual Capacitance (Maximum nf/100 m)				
	6.6	5.6	5.6	5.6

EIA/TIA 568-B HORIZONTAL CABLES *(cont'd)*

	Cat. 3	Cat. 5 Annex N	Cat. 5e	Cat. 6 Draft 11
Capacitance Unbalance Pair to Ground (Maximum pF)				
330	330	330	330	330
Patch Cables				
Insertion Loss dB/100 m				
Insertion Loss limits are derived from the formula listed for horizontal cable increased by 20%				
0.772 MHz	2.7	2.2	2.2	2.2
1.0 MHz	3.1	2.4	2.4	2.4
4.0 MHz	6.7	4.9	4.9	4.6
8.0 MHz	10.2	6.9	6.9	6.4
10.0 MHz	11.7	7.8	7.8	7.2
16.0 MHz	15.7	9.9	9.9	9.1
20.0 MHz		11.1	11.1	10.2
25.0 MHz		12.5	12.5	11.4

EIA/TIA 568-B HORIZONTAL CABLES *(cont'd)*

	Cat. 3	Cat. 5 Annex N		Cat. 5e	Cat. 6 Draft 11
31.25 MHz		14.1		14.1	12.8
62.5 MHz		20.4		20.4	18.5
100.0 MHz		26.4		26.4	23.8
200.0 MHz					34.8
250.0 MHz					39.4
Structural Return Loss (SRL), dB and Return Loss (RL), dB					
	SRL (Min)	SRL (Min)	RL (Min)	RL (Min)	RL (Min)
1 - 10 MHz	12	23		20 +5log(f)	20 + 5log(f)
10 - 16 MHz	12	23			
10 - 20 MHz	12 - 10log(f/10)			25	25
16 - 20 MHz		23			
20 - 100 MHz		16 - 10log(f/100)		25 - 8.6 log(f/20)	
20 - 250 MHz					25 - 8.6log(f/20)

STANDARD MATERIALS FOR STRUCTURED CABLING (568) SYSTEMS

Four-pair 100 ohm UTP cables. The cable consists of 24 AWG thermoplastic insulated conductors formed into four individually twisted pairs and enclosed by a thermoplastic jacket. Four-pair, 22 AWG cables which meet the transmission requirements may also be used. Four-pair, *shielded* twisted pair cables which meet the transmission requirements may also be used.

The diameter over the insulation shall be 1.22 mm (0.048 in) max.

The pair twists of any pair shall not be exactly the same as any other pair. The pair twist lengths shall be selected by the manufacturer to assure compliance with the crosstalk requirements of this standard.

Color Codes.

Pair 1	White-Blue (W-BL)	Blue (BL)
Pair 2	White-Orange (W-O)	Orange (O)
Pair 3	White-Green (W-G)	Green (G)
Pair 4	White-Brown (W-BR)	Brown (BR)

CABLE SPECIFICATIONS

- The diameter of the completed cable shall be less than 6.35 mm (0.25 in)
- The ultimate breaking strength of the completed cable is 90 lb minimum. Maximum *pulling* tension should not exceed 25 lb to avoid stretching.
- The cable tested shall withstand a bend radius of 25.4 mm (1 in) at a temperature of -20˚C without jacket or insulation cracking
- The resistance of any conductor shall not exceed 28.6 ohms per 305 m (1000 ft) at or corrected to a temperature of 20˚C.
- The resistance unbalance between the two conductors of any pair shall not exceed 5% when measured at or corrected to a temperature of 20˚C.
- The mutual capacitance of any pair at 1 kHz shall not exceed 20 nF per 305 M (1000 ft).
- The mutual capacitance of any pair at 1 kHz and measured at or corrected at a temperature of 20˚C, shall not exceed 17 nF per 305 m (1000 ft) for category 4 and category 5 cables.
- The capacitance unbalance to ground at 1 kHz of any pair shall not exceed 1000 pF per 305m (1000 ft).

STRUCTURED CABLING STANDARDS

ANSI coordinates standards activities in the US.EIA/TIA is the most active organization in data and voice cabling standards and has the support of all the major vendors of cabling products. EIA/TIA "568" is probably the most quoted standard in communication cabling.

The TR41.8 committees have developed the following standards relating to building cabling networks:

EIA/TIA 568A Commercial Building Telecommunications Wiring Standard (rev B soon)
EIA/TIA 569A Telecommunications Wiring Pathways and Spaces
EIA/TIA 570 Light Commercial and Residential Telecommunications Cabling
EIA/TIA 606 Telecommunications Cabling System Administration
EIA/TIA 607 Telecommunications System Grounding and Bonding Requirements

Several technical service bulletins (TSBs) have also been published relating to this standard, to clarify various points in the standard:

TSB-36 UTP Categories 3, 4, and 5 Defined
TSB-40A UTP Connecting hardware for Category 3, 4, 5
TSB-53 Additional specifications for STP (shielded twisted pair) hardware
TSB-67 Transmission performance specification for field testing UTP network cabling.
TIA/TSB-72: Centralized Optical Fiber Cabling Guidelines
TSB-75 Defines "zone distribution systems" for horizontal wiring

UTP CABLE PAIRS USED BY TYPICAL NETWORKS

Network Conductor Use			
Network	**568A Pairs**	**568B Pairs**	**Pins Used**
10Base-T	3, 2	2, 3	1-2, 3-6
Token Ring	1, 2	1, 3	4-5, 3-6
TP-PMD (FDDI)	3, 4	2, 4	1-2, 7-8
ATM	3, 4	2, 4	1-2, 7-8
100Base-TX	3, 2	2, 3	1-2, 3-6
100Base-T4	1-4	1-4	1-2, 3-6, 4-5, 7-8
100VG-AnyLAN	1-4	1-4	1-2, 3-6, 4-5, 7-8
Gigabit Ethernet or 1000 Base-T	1-4	1-4	1-2, 3-6, 4-5, 7-8

ETHERNET 10BASE-T STRAIGHT THRU PATCH CORD

		RJ45 Plug	RJ45 Plug	
T2	1	White/Orange	1	TxData + pair 2/—
R2	2	Orange	2	TxData –
T3	3	White/Green	3	RecvData +
R1	4	Blue.	4	pair3
T1	5	White/Blue	5	
R3	6	Green	6	RecvData –
T4	7	White/Brown	7	
R4	8	Brown	8	

ETHERNET 10BASE-T CROSSOVER PATCH CORD

- This cable is used to cascade hubs, or for connecting two Ethernet stations back-to-back without a hub. Note pin numbering of straight-thru patch cord.

RJ45 Plug		RJ45 Plug	
1	Tx+	Rx+	3
2	TX	Rx–	6
3	RX+	Tx+	1
6	Rx+	Tx–	2

DIGITAL PATCH CABLE (DPC) CODING

Pair 1: Green & Red

Pair 2: Yellow & Black

Pair 3: Blue & Orange

Pair 4: Brown & Gray

TELECOMMUNICATIONS OUTLET SPECIFICATIONS

- **100-ohm UTP Cable**—Each four-pair cable shall be terminated in an eight-position modular jack in the work area. The 100-ohm UTP telecommunications outlet shall meet the requirements described in EIA/TIA-570
- **150 ohm STP Cable**—The telecommunications connector used for terminating the 150-ohm STP cable shall be that specified by ANSI/IEEE 802.5 for the media interface connector. This connector shall be designed so that like units will mate when oriented 180 degrees with respect to each other.

STANDARD NETWORKING CONFIGURATIONS

ATM 155 Mbps uses pairs 2 and 4 (pins 1-2, 7-8)

Ethernet 10Base-T uses pairs 2 and 3 (pins 1-2, 3-6)

Ethernet 100Base-T4 uses pairs 2 and 3 (4T+) (pins 1-2, 3-6)

Ethernet 100 Base-T8 uses pairs 1,2,3 and 4 (pins 4-5, 1-2, 3-6, 7-8)

Token-Ring uses pairs 1 and 3 (pins 4-5, 3-6)

TP-PMD uses pairs 2 and 4 (pins 1-2, 7-8)

100VG-AnyLAN uses pairs 1,2,3 and 4 (pins 4-5, 1-2, 3-6, 7-8)

IEEE ETHERNET STANDARDS

- 802.3—Hardware standards for Ethernet cards and cables
- 802.5—Hardware standards for Token Ring cards and cables
- 802.2—The new message format for data on any LAN

ETHERNET FAILURES

Ethernets fail in three common ways:

1. A nail or other object can break one of the conductors.
2. A screw or other object can touch one or more of the conductors and short them to an external grounded metal shield, conduit, or other grounded metal.
3. A station on the network can break down and start to generate a continuous stream of electronic noise, thus blocking legitimate transmissions.

COMMON ETHERNET SYSTEMS

10BASE-5 or Thick Ethernet

- 10BASE-5 is the original Ethernet system. It employs a quarter of an inch diameter, 50 ohm coax cable (with a minimum bend radius of 10 inches). 10BASE-5 segments can run in length up to 500 meters with as many as 100 transceiver connections spaced at least 2.75 yards apart.
- 10BASE-5 transceivers access the media by piercing the thick coaxial cable. These transceiver taps are known as *vampire taps.* Since they don't actually require breaking the physical cable, the electrical signals over the cable are typically fairly clean.
- 10BASE-5 systems were originally envisioned to be cheap and fairly easy to build. The large cable needed simply to be run by rooms where computing equipment would be located. Taps would be made into the cable by using external transceivers. As it turned out, the requirement of an external transceiver and the thick cable, which was expensive and difficult to work with, limited the use of 10BASE-5.

10BASE-2 (THIN ETHERNET AND CHEAPERNET)

- Thin Ethernet was a fairly popular specification and is still used in many environments today. With a maximum segment length of 203.5 yards, it requires that the 50 ohm cable be only .2 inches thick (a bend radius of two inches). It also uses standard BNC connectors and "T's" to provide access to the media. Typically, T's are connected directly to the back of network interface cards, thus eliminating the need for an external transceiver.
- A maximum of 30 transceivers may be inserted onto a Thin Ethernet segment and must be spaced at least 20 inches apart. 3Com hardware is able to handle slightly longer segments, up to 220 yards in length. Unfortunately, mixing other vendor's equipment into an environment where cable runs exceed 203.5 yards can cause problems. For this reason, keeping total lengths to 203.5 yards is recommended.

WIRING SCHEDULES

COLOR CODE	TELCO 25 PAIR	USOC (4 WIRE)	USOC (6 WIRE)	USOC (8 WIRE)
WHT/BLU	26	4	4	5
BLU/WHT	1	3	3	4
WHT/ORG	27	2	2	3
ORG/WHT	2	5	5	6
WHT/GRN	28	4	1	2
GRN/WHT	3	3	6	7
WHT/BRN	29	2	4	8
BRN/WHT	4	5	3	1
WHT/SLT	30	4	2	5
SLT/WHT	5	3	5	4
RED/BLU	31	2	1	3
BLU/RED	6	5	6	6
RED/ORG	32	4	4	2
ORG/RED	7	3	3	7
RED/GRN	33	2	2	8
GRN/RED	8	5	5	1
RED/BRN	34	4	1	5
BRN/RED	9	3	6	4
RED/SLT	35	2	4	3
SLT/RED	10	5	3	6
BLK/BLU	36	4	2	2
BLU/BLK	11	3	5	7
BLK/ORG	37	2	1	8
ORG/BLK	12	5	6	1
BLK/GRN	38	4	4	5
GRN/BLK	13	3	3	4

WIRING SCHEDULES *(cont'd)*

COLOR CODE	TELCO 25 PAIR	USOC (4 WIRE)	USOC (6 WIRE)	USOC (8 WIRE)
BLK/BRN	39	2	2	3
BRN/BLK	14	5	5	6
BLK/SLT	40	4	1	2
SLT/BLK	15	3	6	7
YEL/BLU	41	2	4	8
BLU/YEL	16	5	3	1
YEL/ORG	42	4	2	5
ORG/YEL	17	3	5	4
YEL/GRN	43	2	1	3
GRN/YEL	18	5	6	6
YEL/BRN	44	4	4	2
BRN/YEL	19	3	3	7
YEL/SLT	45	2	2	8
SLT/YEL	20	5	5	1
VLT/BLU	46	4	1	5
BLU/VLT	21	3	6	4
VLT/ORG	47	2	4	3
ORG/VLT	22	5	3	6
VLT/GRN	48	4	2	2
GRN/VLT	23	3	5	7
VLT/BRN	49	2	1	8
BRN/VLT	24	5	6	1
VLT/SLT	50	-	-	-
SLT/VLT	25	-	-	-

WIRING SCHEDULES *(cont'd)*

COLOR CODE	356A	EIA 568B (258A)	10 BASE-T	EIA 568A
WHT/BLU	5	5	1	5
BLU/WHT	4	4	2	4
WHT/ORG	1	1	3	3
ORG/WHT	2	2	6	6
WHT/GRN	3	3	1	1
GRN/WHT	6	6	2	2
WHT/BRN	5	7	3	7
BRN/WHT	4	8	6	8
WHT/SLT	1	5	1	5
SLT/WHT	2	4	2	4
RED/BLU	3	1	3	3
BLU/RED	6	2	6	6
RED/ORG	5	3	1	1
ORG/RED	4	6	2	2
RED/GRN	1	7	3	7
GRN/RED	2	8	6	8
RED/BRN	3	5	1	5
BRN/RED	6	4	2	4
RED/SLT	5	1	3	3
SLT/RED	4	2	6	6
BLK/BLU	1	3	1	1
BLU/BLK	2	6	2	2
BLK/ORG	3	7	3	7
ORG/BLK	6	8	6	8
BLK/GRN	5	5	1	5
GRN/BLK	4	4	2	4

WIRING SCHEDULES *(cont'd)*

COLOR CODE	356A	EIA 568B (258A)	10 BASE-T	EIA 568A
BLK/BRN	1	1	3	3
BRN/BLK	2	2	6	6
BLK/SLT	3	3	1	1
SLT/BLK	6	6	2	2
YEL/BLU	5	7	3	7
BLU/YEL	4	8	6	8
YEL/ORG	1	5	1	5
ORG/YEL	2	4	2	4
YEL/GRN	3	1	3	3
GRN/YEL	6	2	6	6
YEL/BRN	5	3	1	1
BRN/YEL	4	6	2	2
YEL/SLT	1	7	3	7
SLT/YEL	2	8	6	8
VLT/BLU	3	5	1	5
BLU/VLT	6	4	2	4
VLT/ORG	5	1	3	3
ORG/VLT	4	2	6	6
VLT/GRN	1	3	1	1
GRN/VLT	2	6	2	2
VLT/BRN	3	7	3	7
BRN/VLT	6	8	6	8
VLT/SLT	-	-	-	-
SLT/VLT	-	-	-	-

MODULAR PLUG ORIENTATIONS

(8 wire) MOD Plug straight through wiring
1-1 2-2 3-3 4-4 5-5 6-6 7-7 8-8

RJ11

RJ12

MMP

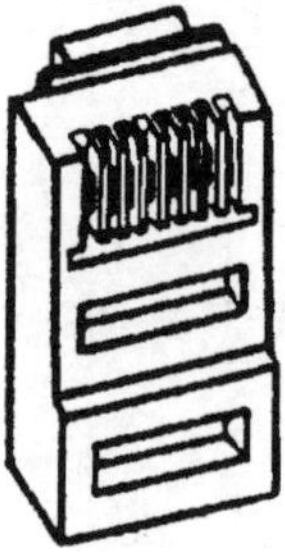
RJ45

(8 wire) MOD Plug crossover wiring
1-8 2-7 3-6 4-5 5-4 6-3 7-2 8-1

MODULAR JACK ORIENTATIONS

10BT crossover wiring
Pair 1: 1-3, 2-6
Pair 2: 3-1, 6-2

CDDI wiring
pair 1: 1-7, 2-8
pair 2: 7-1, 8-2

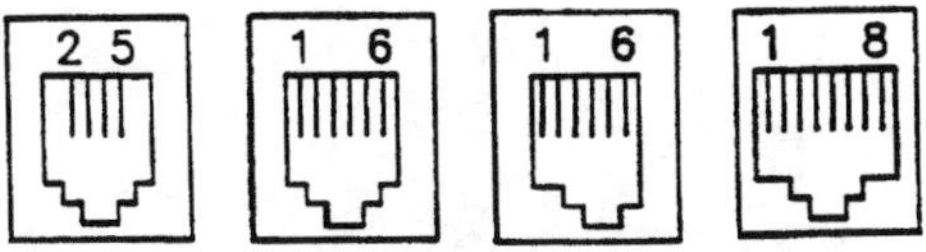

PIN ORIENTATIONS

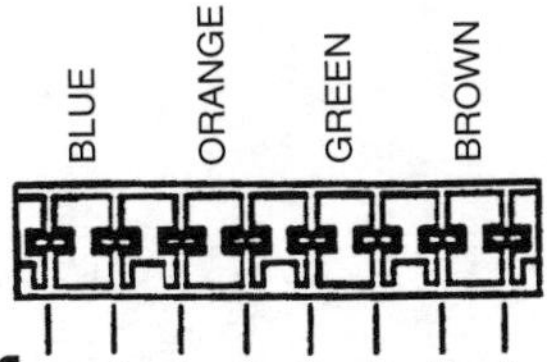

RJ45 PIN NUMBERS									
	5	4	3	6	2	7	8	1	USOC
	5	4	1	2	3	6	7	8	AT&T 258A
	5	4	3	6	1	2	7	8	EIA T-568A

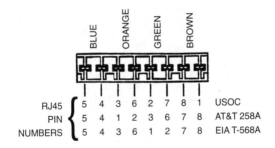

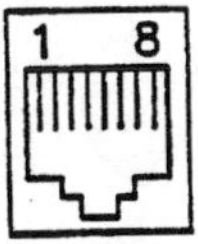

EIA/TIA 568A MODULAR CONNECTOR WIRING

Pin Number	Pair Number	Color Codes
1	3-Tip	W/G
2	3-Ring	G
3	2-Tip	W/O
4	1-Ring	BL
5	1-Tip	W/BL
6	2-Ring	O
7	4-Tip	W/Br
8	4-Ring	BR

EIA/TIA 568B MODULAR CONNECTOR WIRING

Pin Number	Pair Number	Color Codes
1	2-Tip	W/O
2	2-Ring	O
3	3-Tip	W/G
4	1-Ring	BL
5	1-Tip	W/BL
6	3-Ring	G
7	4-Tip	W/BR
8	4-Ring	BR

OPEN DEC CONNECT WIRING

Pin Number	Pair Number	Color Codes
1	3-Tip	W/G
2	3-Ring	G
3	2-Tip	W/O
4	NC	
5	NC	
6	2-Ring	O
7	4-Tip	W/BR
8	4-Ring	BR

USOC MODULAR CONNECTOR WIRING

Pin Number	Pair Number	Color Codes
1	4-Ring	BR
2	3-Tip	W/G
3	2-Tip	W/O
4	1-Ring	BL
5	1-Tip	W/BL
6	2-Ring	O
7	3-Ring	G
8	4-Tip	W/BR

CATEGORY DESIGNATIONS OF UTP CABLES

Category 1: *POTS*—**P**lain **O**ld **T**elephone **S**ervice.

Category 2: Low speed computer terminal and network (ARCNET) applications.

Category 3: Ethernet and 4/16 MB/s Token Ring cabling.

Category 4: For passive 16 MB/s Token Ring

Category 5: For the copper wire versions of FDDI (Fiber Distributed Digital Interface) at 100 *MB/s* (megabits per second).

Category 5e Includes delay skew and powersum NEXT testing.

Category 6 expands the frequency range to 200 MHZ and tightens powersum NEXT specifications

Category 7 is expected to expand the frequency range to 600 MHZ and has even tighter powersum NEXT specifications.

IBM TECHNICAL INTERFACE SPECIFICATION GA27-3773-1 REQUIREMENTS

Standard Specification LAN Cable					
	Type 1&2	**Type 6 & 9**	**Type 8**	**Type 2 VGM**	**Type 2 VGM-DGM**
DC Resistance, Ω/km	57.1	151	148	57.1	-
DC Resistance Unbalance, pF/km	4	4	4	4	-
Pair-Ground Capacitance Unbalance, pF/km	1500	1500	1500	2625	-
Pair-Pair Capacitance Unbalance, pF/km	-	-	-	181	-
Characteristic Impedance, ohms					
1 kHz	-	-	-	600 ± 15%	-
9.6	270 ± 10%	390 ± 10%	390 ± 10%	-	-
10	-	-	-	188 ± 15%	-
38.4	185 ± 10%	235 ± 15%	235% ± 10%	-	-
40	-	-	-	125 ± 15%	-
256	-	-	-	105 ± 15%	-
4.0 MHz	150 ± 10%	150 ± 10%	150 ± 10%	100 ± 25%	-
3 - 20 MHz	150 ± 10%	150 ± 10%	150 ± 10%	-	-
Attenuation, dB/km					
1 kHz	-	-	-	1.28	-
9.6	3	6	6	-	-
38.4	5	7.4	7	-	-
150	-	-	-	7.13	-
772	-	-	-	16.82	-
4.0 MHz	22	33	44	-	-
16 MHz	45	66	88	-	-
Near-End Crosstalk, dB					
9.6 kHz	80	80	70	-	-
38.4	75	75	70	-	-
150	-	-	-	63 (FEXT)	-
772	-	-	-	56	-
3 - 5 MHz	58	52	58	33	80
12 - 20 MHz	40	34	40	-	0

IBM TECHNICAL INTERFACE SPECIFICATION GA27-3773-1 REQUIREMENTS *(cont'd)*

Enhanced Specification LAN Cable

Enhanced Specification LAN Cables must meet the Standard LAN Cable requirements with the exception and/or additions listed below

	Type 1A & 2A	Type 6A	Type 9A
Pair-Ground Capacitance Unbalance, pF/km	1000	1000	1000
	Attenuation, dB		
4 - 20 MHz	$\leq$ 22 sqrt(f/4)		
20 - 600 MHz	$\leq k_O$ sqrt(f/f_O) for $f_L \leq f \leq f_H$ where;		
Balanced Mode:			
k_O =	9.75	14.75	14.75
f_O =	62.5	62.5	62.5
f_L =	20.0	20.0	20.0
f_H =	300	300	300
Common Mode:			
k_O =	9.50	-	13.50
f_O =	50.0	-	50.0
f_L =	50.0	-	50.0
f_H =	600	-	600
	Near-End Crosstalk, dB		
5-300 MHz	$\leq$-58+15log(f/5)	$\leq$-58+15log(f/5)	$\leq$-58+15log(f/5)

POLARIZATION AND SEQUENCE

- **Polarization** is defined as the physical form factor of a modular jack interface. Examples of this include the Western Electric Co 8 wire (WE8W or RJ45) and the Modified Modular Jack (MMJ). If the polarization of equipment does not match the building cabling interface (wall plate, etc.) a mechanical adapter may be used to convert. The male is referred to as a plug and the female as a jack.
- **Sequence** is defined as the order in which the incoming tip/ring pairs are terminated into the modular interface pins, i.e. which pins are pair #1. Examples of sequence are USOC, ATA&T 258A, DEC, and EIA. Sequence is extremely important as a miss-match can result in high levels of noise and cross-talk from unpairing of signals.

POLARIZATION OPTIONS

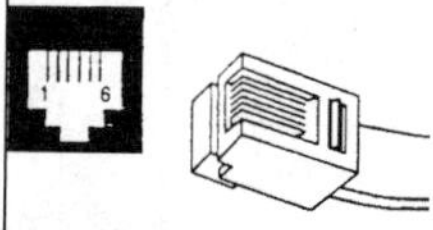

Generically referred to as "RJ11" and "RJ12", this was the modular polarization specified as the standard voice interface by the US telephone companies. Both form factors are identical; the RJ11 interface simply has the outermost two contacts (1 and 6) unpopulated. The modular jack has a center locking tab with a shoulder to hold it into the jack. Pins are numbered 1 to 6 in an RJ12 and 2 to 5 in an RJ11.

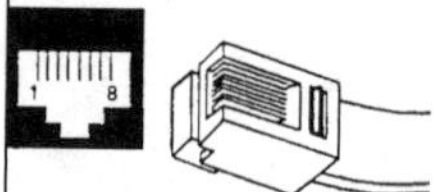

Generically referred to as the "RJ45", this is an 8 conductor version of the above contact. Pins are designated 1 through 8.

The mechanical form factor is wider, not allowing an RJ45 plug to mate with an RJ11 6w jack; the opposite is true, however—an RJ12 plug *will* go into an RJ45 jack resulting in pin 1 on the RJ12 connected to pin 2 of the RJ45, 2 to 3, etc. Pins 1 and 8 of the RJ45 will not be connected through the RJ12, pins 1, 2, 7, and 8 of the RJ45 will not be connected through the RJ11.

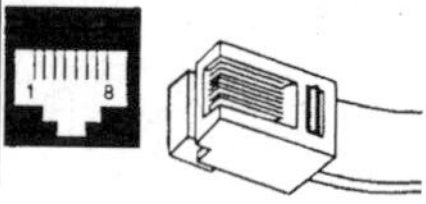

Generically referred to as the "RJ45 Keyed", this was developed as an alternative polarization for specialized telephone lines (balanced). The form factor is the same as the RJ45 with the addition of a tab on the side. This tab prevents a RJ45 plug from going into a RJ45 jack; equipment cords with this polarization therefore could not be connected into a standard RJ45 outlet.

An RJ45 jack will, however, accept an RJ45 plug and provide standard continuity (pin 1 to pin 1, pin 2 to pin 2, etc.). An RJ45 jack will also accept the RJ11/6w plug providing continuity of RJ45k pin 2 to RJ12 pin 1, RJ45K pin 3 to RJ11/6w pin 2, etc. (RJ45K pins 1 and 8 are open circuit).

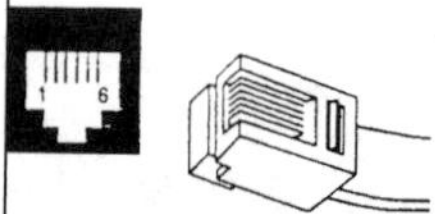

MMJ This polarization was developed by Digital Equipment Corporation (DEC) to create a non-intermatable data jack which could still utilize the tooling, cords, and test gear developed for the telephone jacks. It consists of an RJ12 form factor with the locking tab shifted off to the right. The MMJ will only mate with itself.

SEQUENCE OPTIONS

Sequence is defined as the order in which the incoming pairs are terminated into the modular interface pins. Each pair is designated as a "Tip" conductor and a "Ring" conductor; pair #1 is therefore designated as "T1" and "R1." The sequence defines which pins of the modular interface are defined as T1, R1, T2, R2, etc. Some sequences are only applicable to certain polarizations.

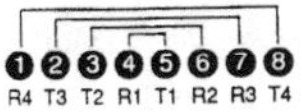

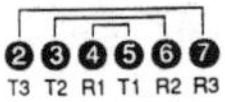

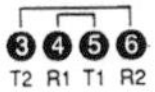

USOC is historically the most common sequence, being used by the US telephone system. Pairs are "nested"; pair #1 is centered, pair #2 is the next two contacts out, etc. This serves to maintain pair-to-pair continuity when, for instance, one pair equipment is connected through a 4 pair circuit. Nesting of the pairs also enables a reversal to be made within each pair through the use of a simple "reversing" line cord (1 to 8, 2 to 7).

USOC is applicable to RJ11, 12, 45, & 45K polarizations. An advantage of the pair nesting of the USOC sequence is that an RJ11/6w jack plugged into an RJ45K plug will work fine as long as quality (correctly toleranced) components are utilized.

MOD-TAP designates the USOC sequence as "Schedule 0" in our product part numbering system.

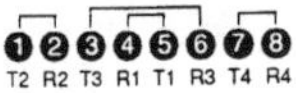

AT&T 258A was widely specified because of the heavy marketing by AT&T of their Premises Distribution System. It is also the specified ISDN sequence and a subset is specified by the IEEE 802.3 10Base-T Ethernet over twisted pair specification. This sequences is only applicable to 8 wire polarizations.

In the 258A sequence, pair 1 corresponds to pair #1 of the USOC sequence, providing backward compatibility with one pair systems (such as analog voice). Pair #2 is where the difference starts, using pins 1 and 2. Pair #3 corresponds to pair #2 of the USOC causing a great deal of confusion, and pair #4 is pins 7 and 8. Due to this structure, a reversing line cord becomes relatively complex.

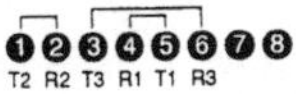

AT&T 356A is a 3 pair version of 258A, leaving pair #4 out (pins 7 and 8 are open).

EIA-232 INTERFACE

DB25 Pin Numbers	Signal/Voltage Source	Signal Designations
14	DTE	Secondary Transmitted Data
1	Common	Shield
15	DCE	(TD) Transmitted Data
2	DTE	(TD) Transmitted Data
16	DCE	Secondary Received Data
3	DCE	(RD) Received Data
17	DCE	Receiver Signal Element Timing
4	DTE	(RTS) Request To Send
18	DTE	Local Loopback
5	DCE	(CTS) Clear To Send
19	DTE	Secondary Request to Send
6	DCE	(DSR) DCE Ready
20	DTE	DTE Ready (DTR)
7	Common	Signal Ground (Common Return)
21	RL=DTE	
	SQ=DCE	Remote Loopback/Sig Quality Det
8	DCE	(DCD) Rcvd Line Signal Detector
22	DCE	Ring Indicator (R)
9	DCE	(+) DC Test Voltage
23	CH=DTE	
10	DCE	(-) DC Test Voltage
24	DTE	Transmit Signal Element Timing
11	--	Unassigned
25	DCE	Test Mode
12	DCE	(SCF/CI) Secondary Rcvd Line Sig Det
13	DCE	Secondary Clear To Send

IBM / AT STYLE RS-232 INTERFACE

	Signal/Voltage Source	Signal Designations
6	DCE	(DSR) Data Set Ready
1	DCE	(DCD) Data Carrier Detect
7	DTE	(RTS) Request to Send
2	DCE	(RD) Received Data
8	DCE	(CTS) Clear to Send
3	DTE	(TD) Transmitted Data
9	DCE	(RI) Ring Indicator
4	DTE	(DTR) Data Terminal Ready
5	Common	(Common Return) Signal Ground

EIA-449 INTERFACE

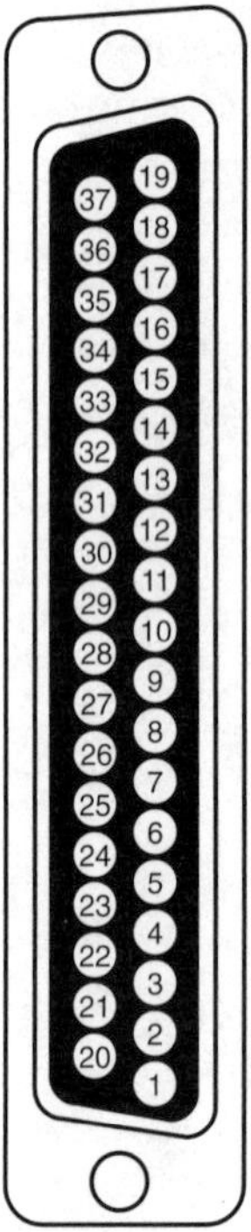

DB37 Pin Numbers	Signal/Voltage Source	Signal Designations
37	Common	Send Common
19	Common	Signal Ground
36	DCE	Standby Indicator
18	DCE	Test Mode (A)
35	Return	Terminal Timing (B)
17	DTE	Terminal Timing (A)
34	DTE	New Signal
16	DTE	Select Frequency
33	DCE	Signal Quality
15	DCE	Incoming Call
32	DTE	Select Standby
14	Remote	Loopback
31	DTE	Return Receiver Ready (B)
13	DCE	Receive Ready (A)
30	Return	Terminal Ready (B)
12	DTE	Terminal Ready (A)
29	Return	Data Mode (B)
11	DCE	Data Mode (A)
28	DTE	Terminal in Service
10	DTE	Local Loopback
27	Return	Clear To Send (B)
9	DCE	Clear To Send (A)
26	Return	Receive Timing (B)
8	DCE	Receive Timing (A)
26	Return	Request to Send (B)
7	DTE	Request To Send (A)
24	Return	Receive Data (B)
6	DCE	Receive Data (A)
23	Return	Send Timing (B)
5	DCE	Send Timing (A)
22	Return	Send Data (B)
4	DTE	Send Data (A)
21	—	Unassigned
3	—	Unassigned
20	Common	Receive Common
2	DCE	Signal Rate Indicator
1	Common	Shield

V.35 INTERFACE

	Signal/Voltage Source	Signal Designations
A	Common	Chassis Ground
B	Common	Ground Signal
C	DTE	Request To Send
D	DCE	Clear to Send
E	DCE	Data Set Read
F	DCE	Data Carrier Detect
H	DTE	Data Terminal Ready
J	DCE	Ring Indicator
K	DTE	Local Test
L	UNASSIGNED	
M	UNASSIGNED	
N	UNASSIGNED	
P	DTE	Send Data (A)
R	DCE	Receive Data (A)
S	DTE	Send Data (B)
T	DCE	Received Data (B)
U	DTE	Terminal Timing (A)
V	DCE	Receive Timing (A)
W	DTE	Terminal Timing (B)
X	DCE	Receive Timing (B)
Y	DCE	Send Timing (A)
Z	UNASSIGNED	
AA	DCE	Send Timing (B)
BB - NN	UNASSIGNED	

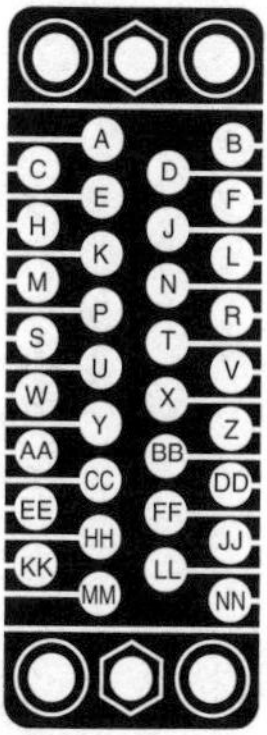

IBM PC KEYBOARD INTERFACE
5 PIN DIN

IBM PC PS/2 KEYBOARD INTERFACE
6 PIN MINI DIN CONNECTOR

APPLE INTERFACE
8 MINI DIN CONNECTOR

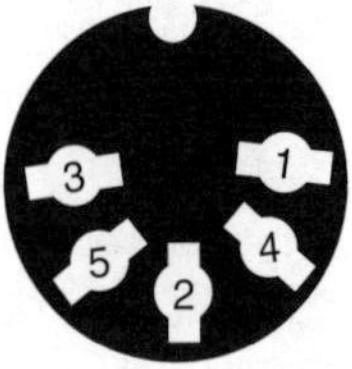

HD15 VGA CONNECTOR

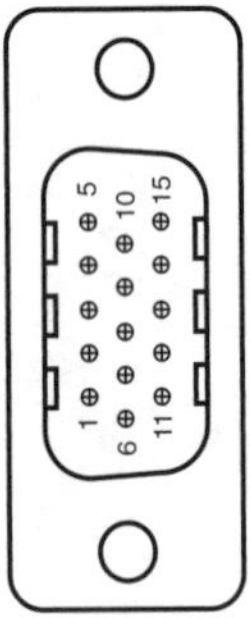

	Signal Designations
1	Red Video
2	Green Video
3	Blue Video
6	Red Return
7	Green Return
8	Blue Return
10	Sync Return
13	Horizontal Return
14	Vertical Sync

COMMON CONNECTOR TYPES

SCSI-26

SCSI-50

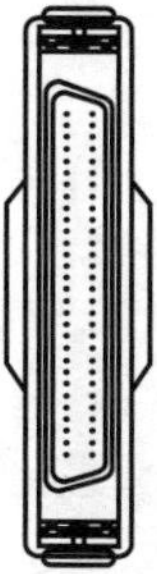

HD-60

HD-26

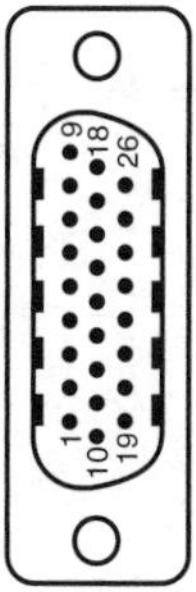

CHAMP-.050

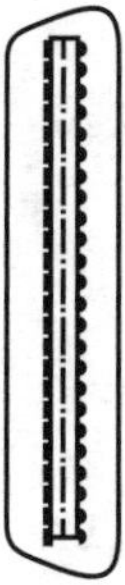

CHAMP-0.8MM

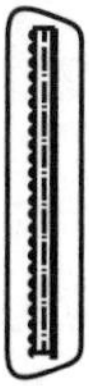

ETHERNET TRANSCEIVER (AUI) INTERFACE

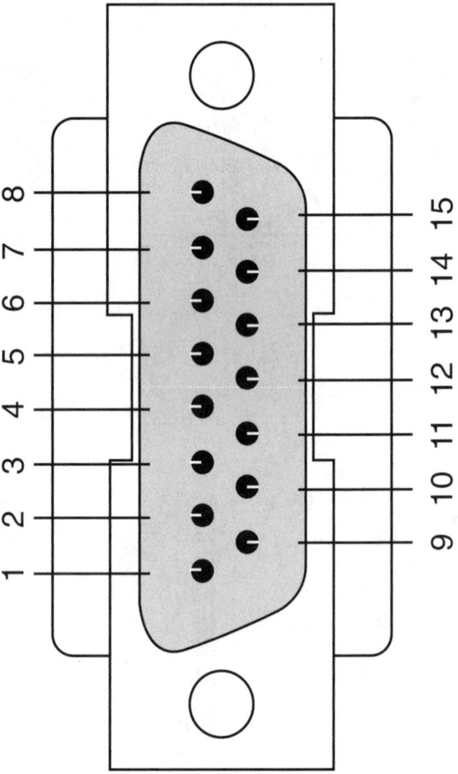

ETHERNET TRANSCEIVER (AUI) INTERFACE *(cont'd)*

Pin	Signal	Pin	Signal
1	Control In (Shield)	9	Control In (Return)
2	Control In	10	Transmit Data (Return)
3	Transmit Data	11	Transmit Data (Shield)
4	Receive Data (Shield)	12	Receive Data (Return)
5	Receive Data	13	Voltage Plus
6	Voltage	14	Voltage (Shield)
7	Control Out	15	Control Out
8	Control Out (Shield)		

10BASE-T/100BASE-T JACKS

RJ-45 Female Modular Jack

1 2 3 4 5 6 7 8

RJ-45 Male Modular Jack

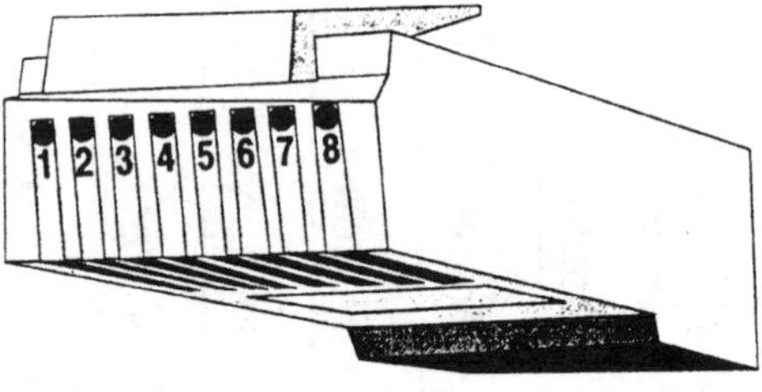

COMMON UNSHIELDED TWISTED PAIR DATA AND VOICE WIRING SCHEMES

8-Wire Jack (10BaseT Data Connections)

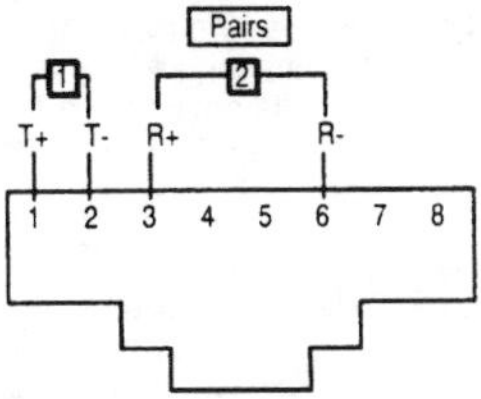

8-Wire Jack (IBM Token Ring Connections)

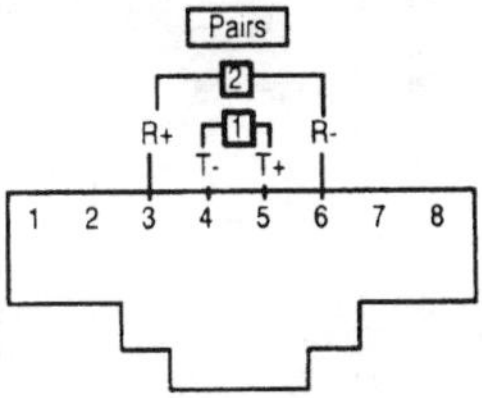

8-Wire Jacks (USOC RJ31X Through RJ37X)

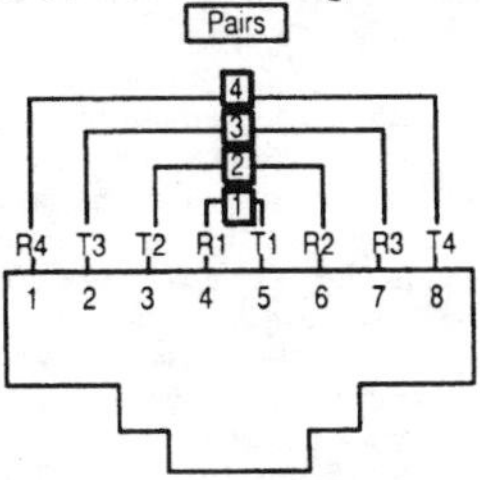

8-Wire Jacks (USOC RJ41 Through RJ48) Also TIA 568B (TIA 568A Swaps Pairs 2 & 3)

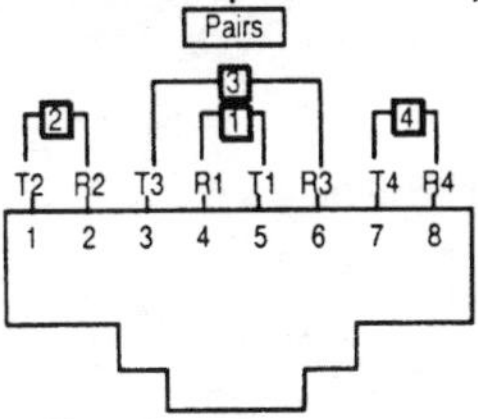

6-Wire Jack (USOC - RJ14W)

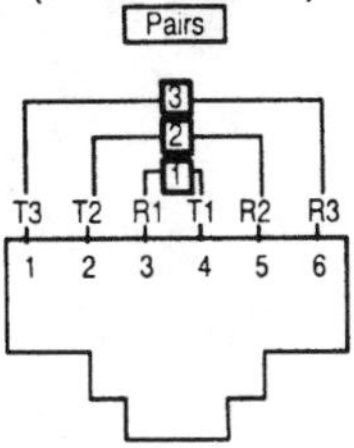

6-Wire Jack Modified Jack (DEC MMJ)

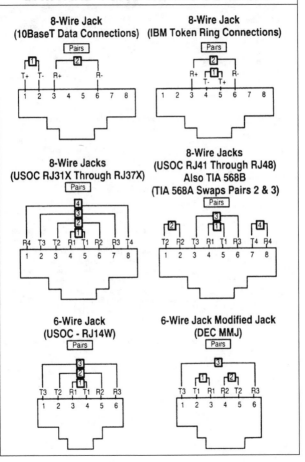

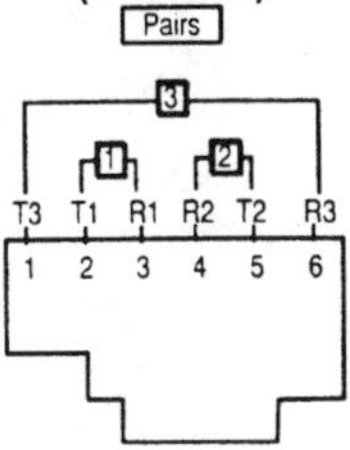

SEQUENCE IDENTIFICATION

Primary	Secondary
White	Blue
Red	Orange
Black	Green
Yellow	Brown
Violet	Slate

- Most data cabling is color coded according to a convention known as "Band Strip Color Coding". This system uses paired colors to identify each pair of conductors. Each paired conductor shares a color pair; one conductor is the primary color with a stripe of the secondary and the other conductor of the pair is the secondary color with a stripe of the primary.
- For a 4 pair cable, the primary color is always white and the secondary colors are blue, orange, green, and brown. Pair #1 is therefore blue / white, pair #2 is orange / white, and so forth. It then becomes simple to define the pairs of the building wiring and locate the modular interface pin each is connected to, either visually or with a continuity tester.
- If the building cabling is not band stripe color coded, the cable jacket must be stripped back far enough to indicate which conductors are twisted with each other into logical pairs. Then, locate which pins of the modular interface each color conductor corresponds to. Next, determine which modular interface pins are paired by referring back to the twisting of the conductors together into pairs. Finally, compare this to the sequence mapping to determine the sequence.

SEQUENCE AND POLARIZATION CONVERSION

Sequence and/or polarization conversion can be accomplished either as individual channels or groups of channels.

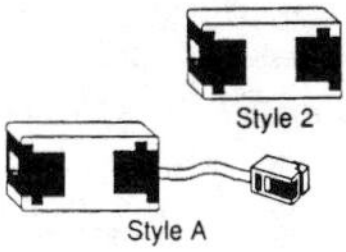

- For single channel conversion, a product called an Adapter is used. This is simply two jacks (style 2) or a jack and a UTP wired plug (style A) wired together to accomplish the proper continuity.

- Multi-channel or group conversion required transition through a 50 position Telco connection. This mass termination has its own sequence where pins 1 through 25 are Ring 1–25 and pins 26 through 50 are Tip 1–25. Products called feeder splits take the 50 position connector (25 pair) and break it out to channel groupings:
  - 12 channels of 2 pair
  - 8 channels of 3 pair
  - 6 channels of 4 pair

Octopus

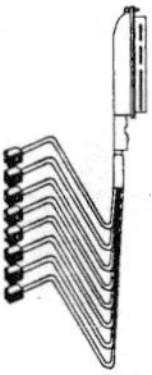

- An Octopus is a feeder split which breaks the 25 pair connector out to plugs. The standard length of the Octopus legs is 2 foot; other lengths available are 4 and 6 foot.

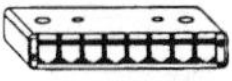

Harmonica

- A Harmonica is a feeder split which breaks the 25 pair connector out to jacks. Harmonicas are also available mounted into a 19" EIA rack panel; this is Modular Patching.

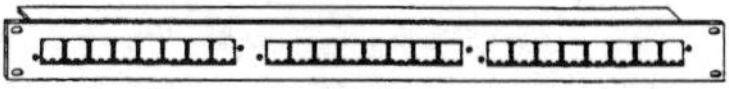

Patching

WIRELESS NETWORKING TERMS

Access Point – A device that transports data between a wireless network and a wired network (infrastructure).

IEEE 802.X – A set of specifications for Local Area Networks (LAN) from The Institute of Electrical and Electronic Engineers (IEEE). Most wired networks conform to 802.3, the specification for CSMA/CD based Ethernet networks or 802.5, the specification for token ring networks. There is an 802.11 committee working on a standard for 1 and 2 Mbps wireless LANs.

Independent Network – A network that provides (usually temporarily) peer-to-peer connectivity without relying on a complete network infrastructure.

Infrastructure Network – A wireless network centered about an access point. In this environment, the access point not only provides communication with the wired network but also mediates wireless network traffic in the immediate neighborhood.

Microcell – A bounded physical space in which a number of wireless devices can communicate. Because it is possible to have overlapping cells as well as isolated cells, the boundaries of the cell are established by some rule or convention.

Multipath – The signal variation caused when radio signals take multiple paths from transmitter to receiver.

Radio Frequency (RF) terms: GHz, MHz, Hz – The international unit for measuring frequency is Hertz (Hz), which is equivalent to the older unit of cycles per second. One Megahertz (MHz) is one million Hertz. One Gigahertz (GHz) is one billion Hertz. The standard US electrical power frequency is 60 Hz, the AM broadcast radio frequency band is 0.55 - 1.6 MHz, the FM broadcast radio frequency band is 88-108 MHz, and microwave ovens typically operate at 2.45 GHz.

Roaming – Movement of a wireless node between two microcells. Roaming usually occurs in infrastructure networks built around multiple access points.

Wireless Node – A user computer with a wireless network interface card (adapter).

EIA 570 RESIDENTIAL NETWORKS

Recognized Cabling by Grade

Cabling	Grade 1	Grade 2
4 pair UTP	1 cable per outlet	2 cables per outlet
	Cat 3 minimum, Cat 5 recommended	Cat 5 Minimum, Cat 5e recommended
75 Ohm Coax	1 cable per outlet	2 cables per outlet
	Series 6	Series 6
Fiber	Not recommended	Optional
		50 or 62.5 micron multimode

Space Allocation for the Distribution Device

Number of Outlets	Grade 1 (height by width)	Grade 2 (height by width)
1 to 8	24" by 16"	36" by 32"
9 to 16	36" by 16"	36" by 32"
17 to 24	48" by 16"	48" by 32"
Greater than 24	60" by 16"	60" by 32"

TYPICAL EIA 570 CABLING SYSTEM

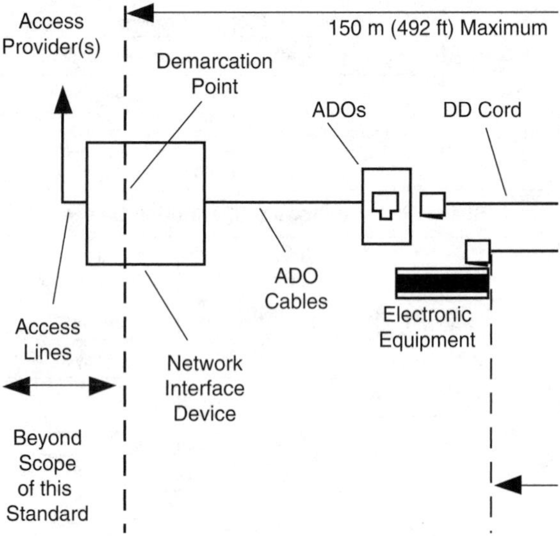

ADO – Auxiliary Disconnect Outlet
DD – Distribution Device

FOR SINGLE RESIDENTIAL NETWORKS

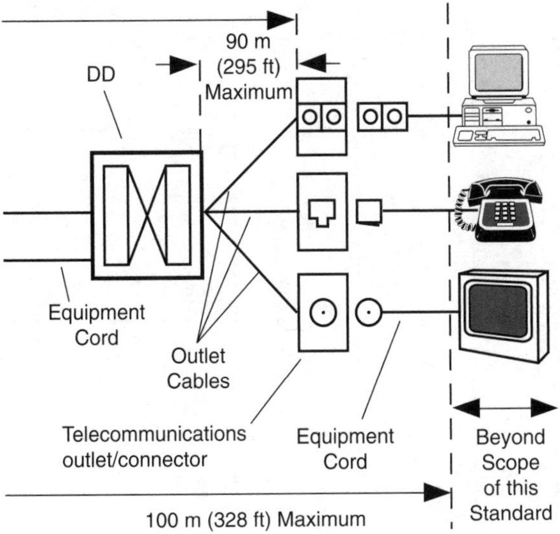

ADO – Auxiliary Disconnect Outlet
DD – Distribution Device

EIA 570 SPACE ALLOCATION GUIDELINES FOR THE DD AND ASSOCIATED EQUIPMENT

Number of Outlet/connectors	Grade 1	Grade 2
1 to 8	410 mm (16 in) wide 610 mm (24 in) high	815 mm (32 in) wide 915 mm (36 in) high
9 to 16	410 mm (16 in) wide 915 mm (36 in) high	815 mm (32 in) wide 915 mm (36 in) high
17 to 24	410 mm (16 in) wide 1220 mm (48 in) high	815 mm (32 in) wide 1220 mm (48 in) high
More Than 24	410 mm (16 in) wide 1525 mm (60 in) high	815 mm (32 in) wide 1525 mm (60 in) high

TYPICAL RESIDENTIAL SERVICES SUPPORTED BY GRADE

Service	Grade 1	Grade 2
Telephone	X	X
Television	X	X
Data	X	X
Multimedia		X

Recognized residential cabling by grade

Cabling	Grade 1	Grade 2
4-Pair UTP	Category 3; Category 5 cable recommended	Category 5; Category 5e cable recommended
75-ohm coax	X	X
Fiber		X (optional)

T568A Eight Position Pin-Pair Assignment

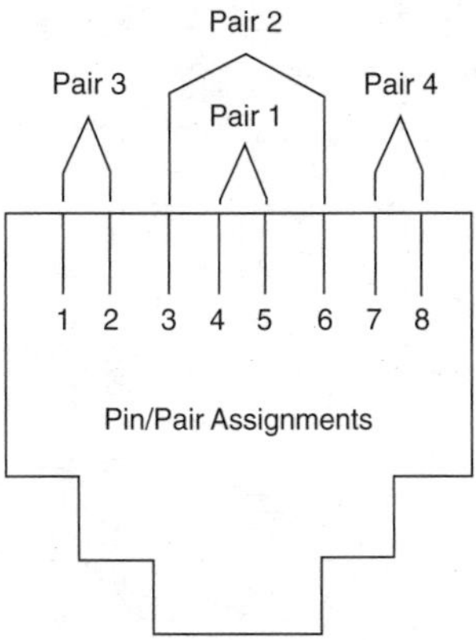

BASIC 570 LINK TEST CONFIGURATION

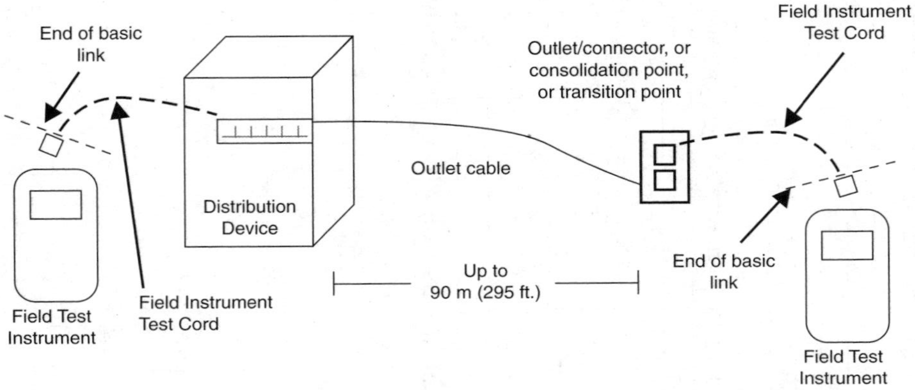

NOTE – Test cords are up to 2 m (79 in.) in length.

EIA 570 CHANNEL TEST CONFIGURATION

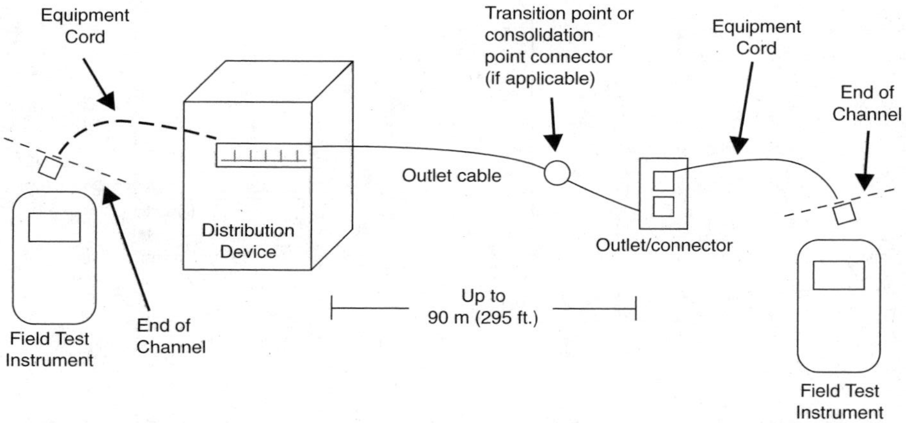

TYPICAL 570 CABLING SYSTEM OF A SINGLE RESIDENTIAL UNIT - FIRST FLOOR

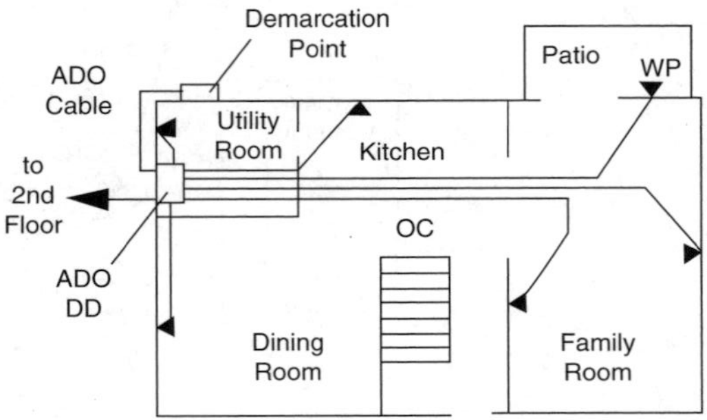

▲ - Telecommunications Outlet/Connector

ADO - Auxiliary Disconnect Outlet

DD - Distribution Device

OC - Outlet Cable

WP - Waterproof Outlet Box

NOTE: Some code bodies limit placing outlets/connectors in bathrooms

TYPICAL 570 CABLING SYSTEM OF A SINGLE RESIDENTIAL UNIT - SECOND FLOOR

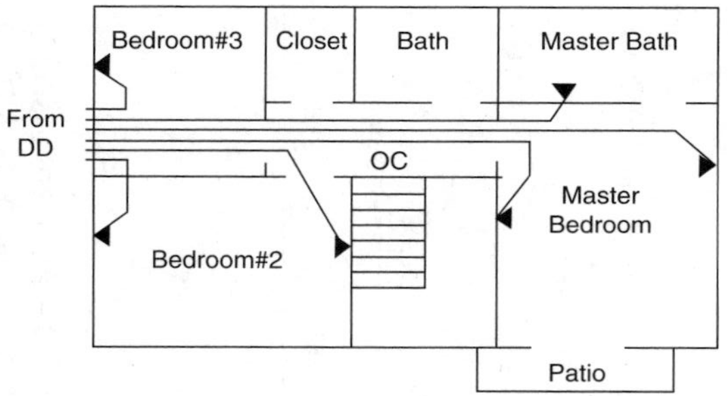

▲ - Telecommunications Outlet/Connector

ADO - Auxiliary Disconnect Outlet

DD - Distribution Device

OC - Outlet Cable

WP - Waterproof Outlet Box

NOTE: Some code bodies limit placing outlets/connectors in bathrooms

TYPICAL 570 CABLING SYSTEM COMPONENTS

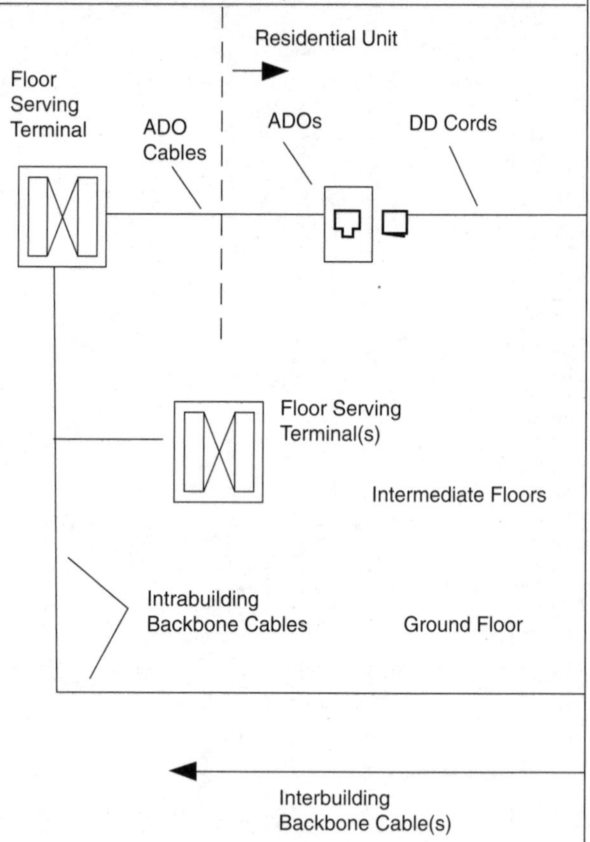

ADO – Auxiliary Disconnect Outlet
DD – Distribution Device

OF A MULTI-TENANT OR CAMPUS ENVIRONMENT

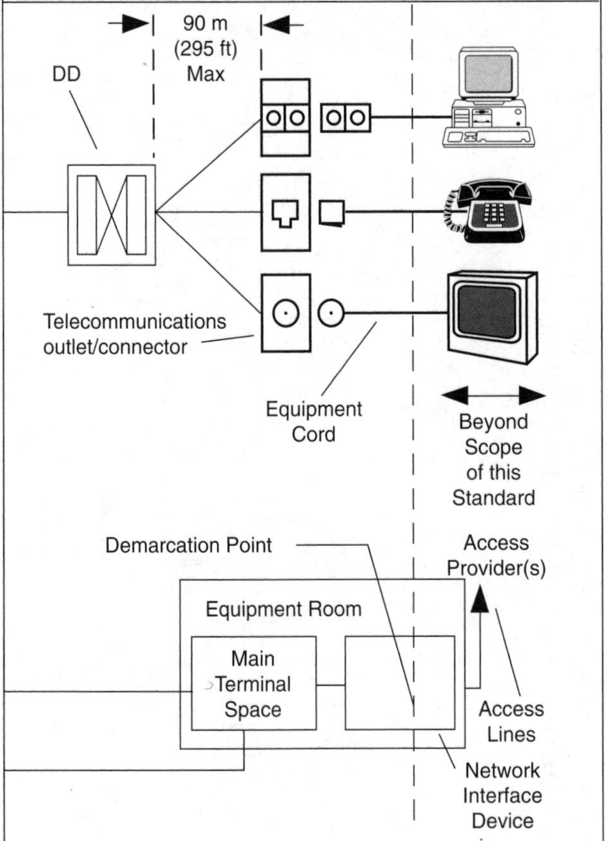

ADO – Auxiliary Disconnect Outlet
DD – Distribution Device

TYPICAL 570 BACKBONE CABLING SYSTEM FOR A MULTI-UNIT BUILDING OR CAMPUS ENVIRONMENT

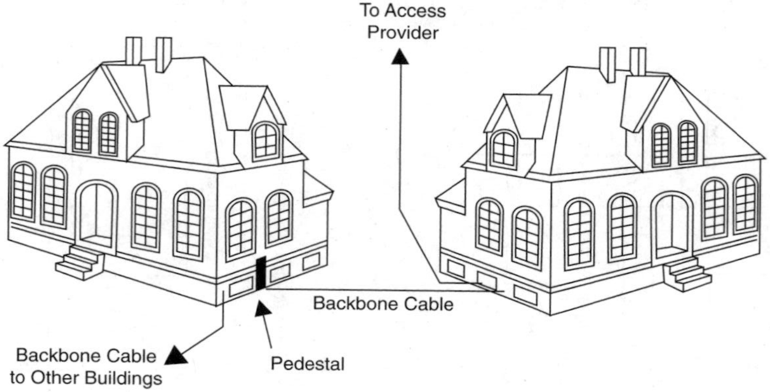

TYPICAL 570 BACKBONE CABLING SYSTEM FOR A MULTI-TENANT BUILDING OR CAMPUS ENVIRONMENT

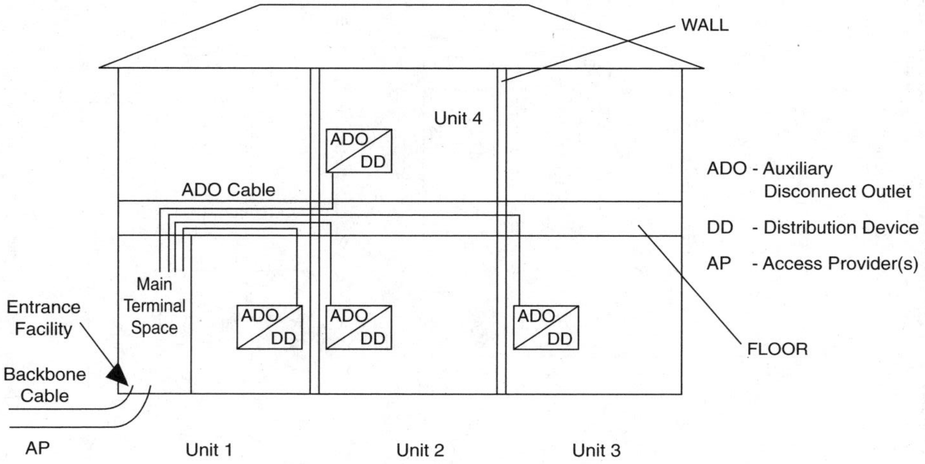

TYPICAL 570 BACKBONE CABLING SYSTEM FOR A

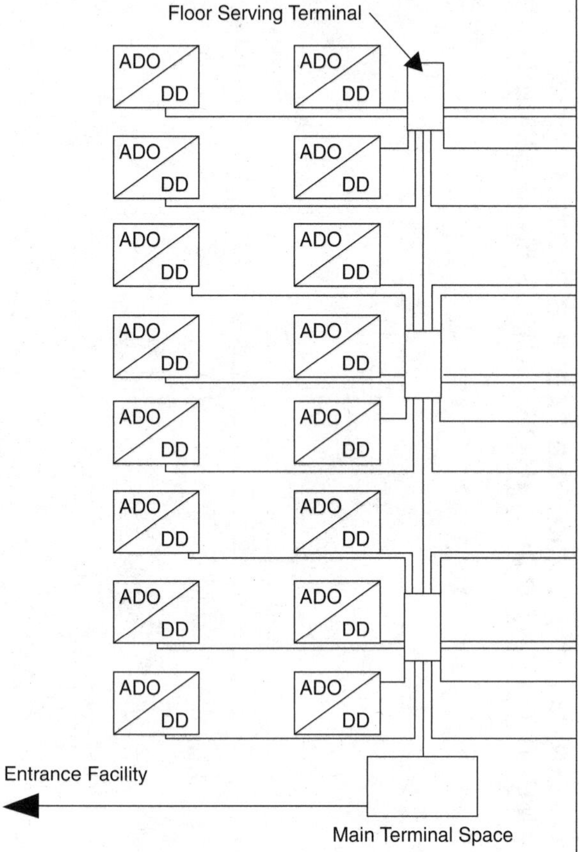

NOTE – Residential units are not similar in design and are not stacked one above the other.

MULTI-FLOOR BUILDING USING FLOOR SERVING TERMINALS

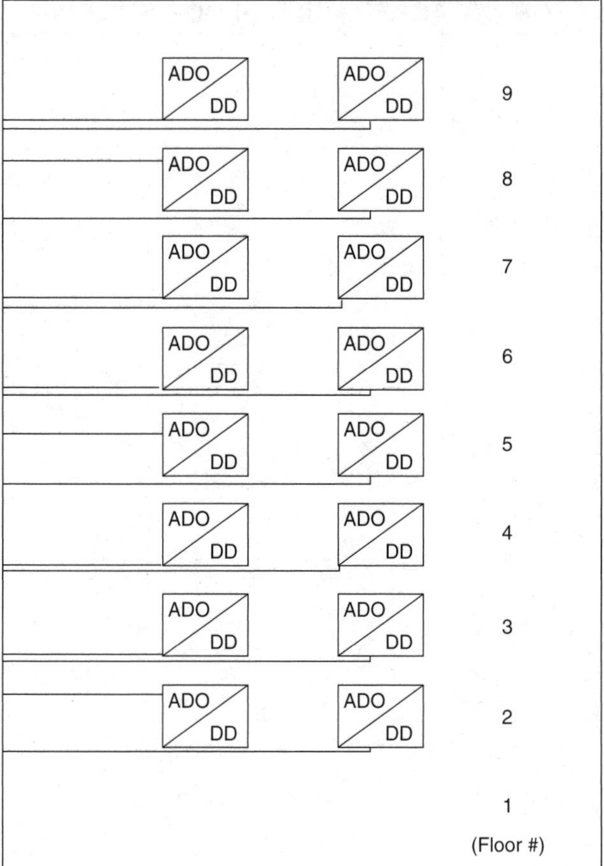

TYPICAL 570 BACKBONE CABLING SYSTEM FOR A STACKED MULTI-FLOOR BUILDING

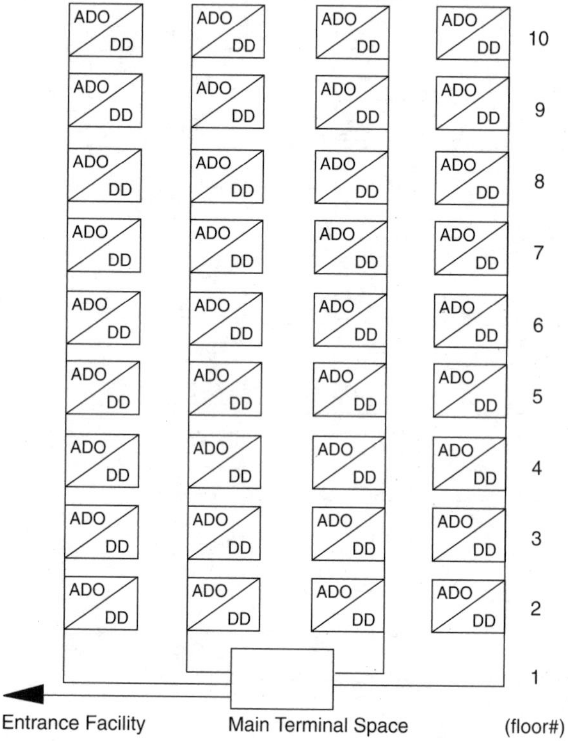

NOTE - All residential units are similar design and stacked one above the other.

CHAPTER 4
DATA CABLES

COMPUTER CIRCUITS

A network is a collection of electrical signaling circuits, each carrying digital signals between pieces of equipment. There are power sources, conductors, and loads involved in the process. The **power source** is a network device that transmits an electrical signal. **The conductors** are the wires that the signal travels over to reach its destination (another network device). The receiver is **the load.** These items, connected together, make up a complete circuit.

In the computer world, the electric signal transmitted by an energy source is a digital signal known as a **pulse.** Pulses are simply the presence of voltage and a lack of the presence of voltage, generated in a sequence. These pulses are used to represent a series of ones and zeros and ones (the presence of voltage being a 1, and the absence of voltage being a 0). These zeros or ones are called *bits.* Many years ago, computer engineers began using groupings of eight bits to represent digital "words," and to this day, a series of 8 bits is called a *byte.* These terms are used everywhere in the computer fields.

The key to successful signal transmission is that when a load receives an electrical signal, the signal must have a voltage level and configuration consistent with what had been originally transmitted by the energy source. If the signal has undergone too much corruption, the load won't be able to interpret it accurately.

A good cable will transfer a signal without too much distortion of the signal, while a bad cable will render a signal useless.

DATA SIGNAL TRANSMISSION

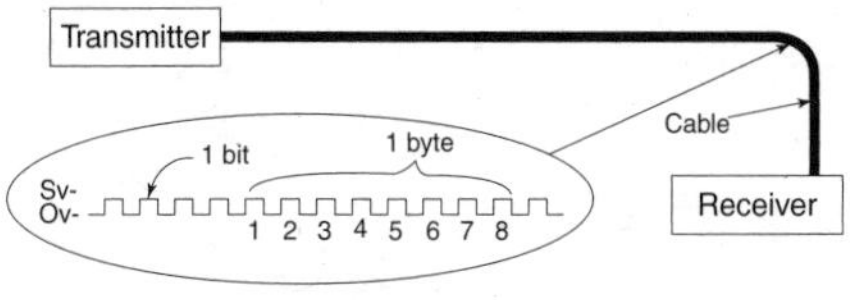

COPPER WIRE LIMITATIONS

- **Due to the electrical properties of copper wiring, data signals will undergo some corruption during their travels.** Signal corruption within certain limits is acceptable, but if the electrical properties of the cable will cause serious distortion of the signal, that cable must be replaced or repaired.

- **As a signal propagates down a length of cable, it loses some of its energy.** So, a signal that starts out with a certain input voltage, will arrive at the load with a reduced voltage level. The amount of signal loss is known as attenuation, which is measured in decibels, or dB. If the voltage drops too much, the signal may no longer be useful.

- **Attenuation has a direct relationship with frequency and cable length.** The higher frequency used by the network, the greater the attenuation. Also, the longer the cable, the more energy a signal loses by the time it reaches the load.

- **A signal loses energy during its travel because of electrical properties at work in the cable.** For example, every conductor offers some dc resistance to a current (sometimes called copper losses). The longer the cable, the more resistance it offers.

- **Resistance reduces the amount of signal passing through the wires—it does not alter the signal.** Reactance, inductive or capacitive, distorts the signal.

- **The two concerns of signal transmission are:**

  1. That enough signal gets through. (Quantity.)
  2. That the signal is not distorted. (Quality.)

ROUGHING AND TRIMMING

Like power wiring, the installation of data cabling consists of two primary phases, roughing in the wiring, then trimming it later.

- **The rough-in phase:** During the rough-in phase, the important things are that all of the cables are put in the proper places, and that they are installed carefully (not bent too tightly, pulled too hard, skinned, or otherwise damaged. At this time, it is also important to consider the routing of the cables, especially if they are unshielded. Unshielded copper cables should never be placed too closely to sources of electromagnetism, such as motor windings, transformers, ballasts, or the like.

- **Fire-stopping:** It is also important to consider fire stopping. Note where the fire barriers in the structure are, and make sure that you make proper allowances for crossing them.

- **Cable Protection:** One area in which roughing data cable differs from roughing-in power wiring is that you must be very sure that your cables will be protected during the construction process (while you are not there). This is critical in situations where there will be a long time between your cable installation and installing the jacks. In such situations, it is up to you to protect your cables any way that will work. If you do not protect them, they may be pulled and twisted by accident. This will damage the cables, even though the damage does not show until you get a tester on them. In other situations, such as when you have a complete raceway system to use, there may be very little time between the cable installation and the wiring of the jacks.

- **Extra Cable:** Also like power wiring, it is important to leave enough extra cable at each outlet point. The recommended lengths are a minimum of 3 meters in the telecommunications closet for both twisted-pair and fiber cable, one meter for fiber and 30 centimeters for twisted-pair cable at the outlet. (Notice that when you move from power wiring to data cabling, the units of measurement switch from English to metric.) Also, remember to check your specifications for requirements on extra cable.

- **Trimming:** Trimming data cabling is pretty much the same as trimming power wiring (strip the cables, install the devices and plates, etc.), except that a lot more testing is required. When trimming power wiring, we generally test by flipping a switch or hitting the outlet with a Wiggy. Either power is present, or it is not. Testing data cabling is not so simple. Remember, you need to test not only for the presence of the signal, but also for the quality of the signal.

CABLE COLORS

Horizontal voice cables	Blue
Inter-building backbone	Brown
Second-level backbone	Gray
Network connections & auxiliary circuits	Green
Demarcation point, telephone cable from Central Office	Orange
First-level backbone	Purple
Key-type telephone systems	Red
Horizontal data cables, computer & PBX equipment	Silver or White
Auxiliary, maintenance & security alarms	Yellow

SEPARATION FROM SOURCES OF INTERFERENCE

Unshielded data cables should not be installed near sources of electromagnetism. There is a standard that specifies these distances for structured data cabling systems. EIA/TIA-569, the cabling pathways standard, specifies the following:

Minimum Separation Distance from Power Source at 480V or less

CONDITION	<2kVA	2-5kVA	>5kVA
Unshielded power lines or electrical equipment in proximity to open or non-metal pathways	5 in.	12 in.	24 in.
Unshielded power lines or electrical equipment in proximity to grounded metal conduit pathway	2.5 in.	6 in.	12 in.
Power lines enclosed in a grounded metal conduit (or equivalent shielding (in proximity to grounded metal conduit pathway	—	6 in.	12 in.
Transformers & electric motors	40 in.	40 in.	40 in.
Fluorescent lighting	12 in.	12 in.	12 in.

MINIMUM BENDING RADII

According to a draft version of EIA-568, the minimum bend radius for UTP is 4 times outside cable diameter, or about one inch. For multi-pair cables the minimum bending radius is 10 × outside diameter. The minimum bend radii for Type 1A Shielded Twisted Pair (100 Mb/s STP) is 7.5 cm (3-in) for non-plenum cable, 15 cm (6-in) for the stiffer plenum-rated kind.

For optical cables not under tension, the minimum bend radius is 10 times diameter; and for cables under tension, no less than 20 times cable diameter. The standard goes on to state that no optical cable will be bent on a radius less than 3.0 cm (1.18-in).

A different standard, ISO DIS 11801 (essentially a parallel standard to the one mentioned above), for 100 ohm and 120 ohm balanced cable lists three different minimum bend radii. Minimum for pulling during installation is 8 times cable diameter, minimum installed radius is 6 times for riser cable, and 4 times cable diameter for horizontal runs. For fiber optic cables, the requirements are the same as those stated above.

Some manufacturers recommendations differ from the above, so it is worth checking the spec sheet for the cable you plan to use.

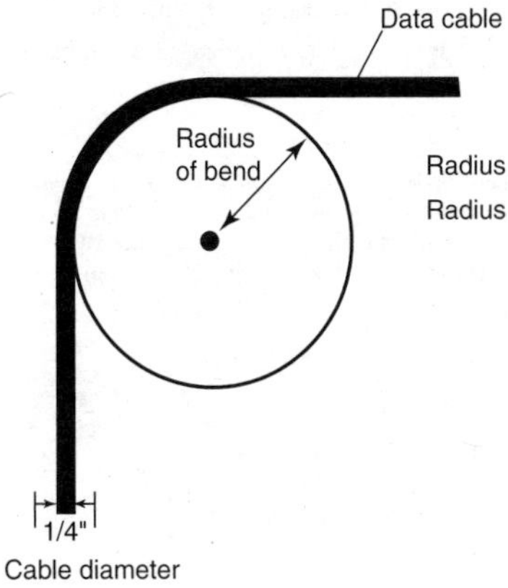

Radius of 10 × diameter = $2^1/_2$"

Radius of 20 × diameter = 5"

COMMON TYPES OF NETWORK CABLING

- **Unshielded twisted pair cables, 22-24 gauge** (UTP)

  Advantages—Inexpensive; may be in place in some places; familiar and simple to install.

  Disadvantages—Subject to interference, both internal and external; limited bandwidth, which translates into slower transmissions. Somewhat vulnerable to security breaches; may become obsolete quickly because of new technologies.

- **Shielded twisted pair cables, 22-24 gauge** (STP)

  Advantages—Easy installation; reasonable cost; resistance to interference; better electrical characteristics than unshielded cables; better data security; easily terminated with modular connector.

  Disadvantages—May become obsolete due to technical advances; can be tapped, breaching security.

- **Coaxial cables**

  Advantages—Familiar and fairly easy to install; better electrical characteristics (lower attenuation and greater bandwidth) than shielded or unshielded cables; highly resistant to interference; generally good data security; easy to connect.

  Disadvantages—May become obsolete due to technological advances; can be tapped, breaching security.

- **Optical fiber cables**

  Advantages—Top performance; excellent bandwidth (high in the gigabit range, and theoretically higher); very long life span; excellent security; allows for very high rates of data transmission; causes no interference and is not subject to electromagnetic interference; smaller and lighter than other cable types.

  Disadvantages—Slightly higher installed cost than twisted-pair cables.

ATTENUATION FOR COAXIAL AND UTP CABLES

Frequency (MHz)	Attenuation (dB/100 meters)				
	Thick Coax	Thin Coax	Cat. 3 UTP	Cat. 4 UTP	Cat. 5 UTP
1	0.62	1.41	2.6	2.2	2.0
10	1.70	4.26	9.7	6.9	6.5
20		6.0		10.0	9.3
50	3.94	9.54			
100		13.70			22.0

Note: UTP figures are based on TIA/EIA requirements for horizontal cable.

GENERAL RULES FOR CABLE INSTALLATIONS

1. Do not exceed a pulling tension of 20% of the ultimate breaking strength of the cable (these figures are available from the cable maker).
2. Lubricate the raceway generously with a suitable pulling compound. (Check with the manufacturer for types of lubricants that are best suited to the type of cable.)
3. Use pulling eyes for manhole installations.
4. For long underground runs, pull the cable both ways from a centrally located manhole to avoid splicing. Use pulling eyes on each end.
5. Do not bend, install, or rack any cable in an arc of less than 12 times the cable diameter.

INSTALLATION REQUIREMENTS

Article 800 of the NEC covers communication circuits, such as telephone systems, computer networks built around telephone-style cables, and outside wiring for fire and burglar alarm systems. Generally these circuits must be separated from power circuits and grounded. In addition, all such circuits must be separated from power circuits and grounded. In addition, all such circuits that run out of doors (even if only partially) must be provided with circuit protectors (surge or voltage suppressors).

Article 725 of the National Electrical Code covers a few types of network cabling. Most Category 5 cables, however, are rated under article 800, rather than under article 725.

The requirements of article 800 are these:

CONDUCTORS ENTERING BUILDINGS

If communications and power conductors are supported by the same pole, or run parallel in span, the following conditions must be met:

1. Wherever possible, communications conductors should be located below power conductors.
2. Communications conductors cannot be connected to crossarms.
3. Power service drops must be separated from communications service drops by at least 12 inches.

Above roofs, communications conductors must have the following clearances:

1. Flat roofs: 8 feet.
2. Garages and other auxiliary buildings: None required.
3. Overhangs, where no more than 4 feet of communications cable will run over the area: 18 inches.
4. Where the roof slope is 4 inches rise for every 12 inches horizontally: 3 feet.

Underground communications conductors must be separated from power conductors in manhole or handholes by brick, concrete, or tile partitions.

Communications conductors should be kept at least 6 feet away from lightning protection system conductors.

INTERIOR COMMUNICATIONS CONDUCTORS

Communications conductors must be kept at least 2 inches away from power or Class 1 conductors, unless they are permanently separated from them or unless the power or Class 1 conductors are enclosed in one of the following:

1. Raceway
2. Type AC, MC, UF, NM, or NM cable, or metal-sheathed cable.

Communications cables are allowed in the same raceway, box, or cable with any of the following:

1. Class 2 and 3 remote-control, signaling, and power-limited circuits.
2. Power-limited fire protective signaling systems.
3. Conductive or nonconductive optical fiber cables.
4. Community antenna television and radio distribution systems.

- Communications conductors are not allowed to be in the same raceway or fitting with power or Class 1 circuits.
- Communications conductors are not allowed to be supported by raceways unless the raceway runs directly to the piece of equipment the communications circuit serves.
- Openings through fire-resistant floors, walls, etc. must be sealed with an appropriate firestopping material.
- Any communications cables used in plenums or environmental air-handling spaces must be listed for such use.

REQUIREMENTS OF ARTICLE 725

Article 725 covers Class 1, 2, and 3 remote control, signaling, and power-limited circuits. This article can be very confusing if you do not understand that circuits designated as Class 1, 2, or 3 can be either *power-limited* circuits, OR *signaling* circuits,

Every circuit covered by this article is Class 1, 2, or 3.

Some of them are power-limited,

and *some* are signaling circuits.

It is important to note that this article does not apply to such circuits that are part of a device or appliance. It applies only to separately installed circuits.

DEFINITIONS

One of the more difficult parts of this article is that it involves a lot of terms with which most of us are not familiar. The key terms are as follows:

- **Class 1:** Circuits that are supplied by a source that has an output of no more than 30 volts (AC or DC) and 1000 volt-amperes. (*Volt-amperes* is essentially the same thing as *Watts*. You will find the term volt-amps used when relating to transformers and similar devices, since it is technically more correct for such uses than Watts.) See section 725-11.

- **Class 2:** Circuits that are inherently limited in capacity. They may require no overcurrent protection, or may have their capacity regulated by a combination of overcurrent protection and their power source. There are a number of voltage, current, and other characteristics that define class 2 circuits. These characteristics are detailed in tables 725-31(a) and 725-31(b).

- **Class 3:** Class 3 circuits are very similar to class 2 circuits. They are inherently limited in capacity. They may require no overcurrent protection, or may have their capacity regulated by a combination of overcurrent protection and their power source. The voltage, current, and other characteristics that define class 3 circuits and differentiate them from 2 circuits are detailed in tables 725-31(a) and 725-31(b).

- **Power-Limited:** This refers to a circuit that has a self-limited power level. That is, whether because of impedance, overcurrent protection, or other power source limitations, these circuits can only operate up to a limited level of power.

- **Remote control and signaling circuits:** These are class 1, 2, or 3 circuits that do not have a limited power rating.

- **Supply side:** This is essentially the same thing as "line side." This term is used instead of "line" because of possible confusion. (Someone might think it referred to the power circuits, rather than simply the power that feeds the circuit.)

- **CL2, CL3P, CL2R, etc.:** These are specific types of power-limited cables. See table 725-50.

- **PLTC:** Power Limited Tray Cable.

CLASS 1 REQUIREMENTS

A remote control, signaling, or power limited circuit is the part of the wiring system between the load side of the overcurrent device or limited power supply and any equipment connected to the circuit.

There are a lot of different requirements listed in article 725. The ones we will cover here will be the most important of the installation requirements. We will begin with class 1 requirements. The requirements are as follows:

Except for transformers, the power supplies that feed class 1 circuits must be protected by an overcurrent device that is rated 167% of the power supply's rated current. This overcurrent device can be built into the power supply; but if so, it can not be interchangeable with devices of a higher rating. In other words, you cannot use any interchangeable fuse (as most are) in this power supply. Transformers feeding these circuits have no requirements except those stated in article 450.

All remote control circuits that could cause a fire hazard must be classified as class 1, even if their other characteristics would classify them as Class 2 or Class 3. This applies also to safety control equipment. (This does not include heating and ventilating equipment.)

Class 1 circuits are not allowed in the same cable with communications circuits.

Power sources for these circuits must have a maximum output (note: this is *maximum output,* not *rated power;* the two terms are very different) of no more than 2,500 VA. This does not apply to transformers.

Class remote control and signaling circuits can operate at up to 600 volts, and their power sources need not be limited.

All conductors in class 1 circuits that are #14 AWG or larger must have overcurrent protection. Derating factors cannot be allowed. #18 conductors must be protected at 7 amps or less, and #16 conductors at 10 amps or less. There are three exceptions to this. See 725-12.

Any required overcurrent devices must be located at the point of supply. See 725-13 for two exceptions.

Class 1 circuits of different sources can share the same raceway or cable, provided that they all have insulation rated as high as the highest voltage present.

If only class 1 conductors are in a raceway, the allowable number of conductors can be no more than easy installation and heat dissipation will allow.

DEFINITIONS OF CLASS 2 AND CLASS 3 CIRCUITS

These are circuits that are inherently limited in capacity. They may require no overcurrent protection, or may have their capacity regulated by a combination of overcurrent protection and their power source. There are a number of voltage, current, and other characteristics that define these circuits. These characteristics are detailed in tables 725-31(a) and 725-31(b).

If you refer to the tables referenced above, you will find that there are quite a few combinations of circuit characteristics that can make the circuit Class 2 or Class 3. Note that the characteristics for AC and DC circuits are different. These tables each have two particular groupings—Circuits that require overcurrent protection, and circuits that do not.

The only substantial differences between Class 2 and Class 3 circuits is that Class 3 circuits generally have higher voltage and power ratings than Class 2 circuits.

CLASS 2 & 3 REQUIREMENTS

The basic requirements for Class 2 and Class 3 circuits are as follow:

Power supplies for Class 2 or 3 circuits may not be connected in parallel unless specifically designed for such use.

The power supplies (when necessary) that feed class 2 or 3 circuits must be protected by an overcurrent device that can not be interchangeable with devices of a higher rating. You cannot use an interchangeable fuse in this power supply. The overcurrent device can be built in to the power supply.

All overcurrent devices must be installed at the point of supply.

Transformers that are supplied by power circuits may not be rated more than 20 amps. They may, however, have #18 AWG leads, so long as the leads are no more than 12 inches long.

Class 2 or 3 conductors must be separated at least 2 inches from all other conductors, except in the following situations:

1. If the other conductors are enclosed in raceway, metal-sheathed cables, metal-clad cables, NM cables, or UF cables.
2. If the conductors are separated by a fixed insulator, such as a porcelain or plastic tube.

Class 2 & 3 REQUIREMENTS *(cont'd)*

Class 2 or 3 conductors may not be installed in the same raceway, cable, enclosure, or cable tray with other conductors, except:

1. If they are separated by a barrier.
2. In enclosures, if Class 1 conductors enter only to connect equipment which is also connected to the Class 2 and/or 3 circuits.
3. In manholes, if the power or Class 1 conductors are in UF or metal-enclosed cable.
4. In manholes, if the conductors are separated by a fixed insulator, in addition to the insulation of the conductors.
5. In manholes, if the conductors are mounted firmly on racks.
6. If the conductors are part of a hybrid cable for a closed loop system. (See article 780.)

When installed in hoistways, Class 2 or 3 conductors must be enclosed in rigid metal conduit, rigid nonmetallic conduit, IMC, or EMT. It can be installed in elevators as allowed by section 620-21.

In shafts, Class 2 or 3 conductors must be kept at least 2 inches away from other conductors.

Any cables that are used for Class 2 or 3 systems must be marked as resistant to the spread of flame.

All Class 2 and 3 cables are marked, listing the areas in which they may be installed. The cables may be installed in the listed areas only.

Two or more Class 2 circuits can be installed in the same enclosure, cable, or raceway, so long as they are all insulated for the highest voltage present.

Two or more Class 3 circuits may share the same raceway, cable, or enclosure.

Class 2 and Class 3 circuits can be installed in the same raceway or enclosure with other circuits, provided the other circuits are in a cable of one of the following types:

Power-limited signaling cables (see article 760)

Optical fiber cables

Communication cables (see article 800)

Community antenna cables (see article 820)

When Class 2 or 3 conductors extend out of a building, and are subject to accidental contact with systems operating at over 300 volts to ground (not 300 volts between conductors—but 300 volts to ground), they must meet all the requirements of section 800-30 for communication circuits.

CABLING CLASSIFICATIONS

Following are the cable classifications which are applicable to data cabling are these:

Article 725, Class 2:

725-38(b)1	CL2X	Class 2 cable, limited use
725-38(b)1	CL2	Class 2 cable
725-38(b)2	CL2R	Class 2 riser cable
725-38(b)3	CL2P	Class 2 plenum cable

Article 800:

800-3(b)1	CMX	Communications cable limited use
800-3(b)1	CM	Communications cable
800-3(b)2	CMR	Communications riser cable
800-3(b)3	CMP	Communications plenum cable

Article 770:

OFNP (Optical Fiber Nonconductive Plenum)
OFNR (Optical Fiber Nonconductive Riser)

Four-pair 100 ohm UTP cables. The cable consists of 24 AWG thermoplastic insulated conductors formed into four individually twisted pairs and enclosed by a thermoplastic jacket. Four-pair, 22 AWG cables which meet the transmission requirements may also be used. Four-pair, *shielded* twisted pair cables which meet the transmission requirements may also be used.

The diameter over the insulation shall be 1.22 mm (0.048 in) max.

The pair twists of any pair shall not be exactly the same as any other pair. The pair twist lengths shall be selected by the manufacturer to assure compliance with the crosstalk requirements of this standard.

Color Codes

Pair 1	White-Blue (W-BL)	Blue (BL)
Pair 2	White-Orange (W-O)	Orange (O)
Pair 3	White-Green (W-G)	Green (G)
Pair 4	White-Brown (W-BR)	Brown (BR)

UNDERGROUND RACEWAYS

- While installing directly-buried cables are certainly common, there are many times when underground cables are installed in raceways. **The advantages in this are superior protection, and the ability to add conductors at a future date** with a minimum of expense. Obviously, it is far more expensive to install a conduit than it is to install a cable.
- When installing underground conduits, it is important to **install more raceway than you need** at the time of installation. This allows you to install more cables in the future. Pulling more cables in a conduit that is already partially-filled is not generally a viable plan—working in pre-filled conduits is frequently more difficult than planned.
- Another thing to do when installing underground conduits is to **install multiple innerducts** inside one large conduit. This allows you to have three or four separate mini-raceways inside of the one large raceway.

  This method often eliminates the need for pull rope and reduces the amount of lubricant required.

BURYING CABLE

- There are a number of reasons why **buried cable is preferred,** rather than placing poles and running the cable aerially (which is frequently a less-expensive option). One reason is that a direct burial cable is virtually free from storm damage and has lower maintenance costs than aerial cable. In addition, aerial installations often lack aesthetic appeal, and in some communities, are even prohibited.
- Although any cable can be buried in the earth, a **cable that is specifically designed for direct burial will have a longer life.** A cable with a high-density polyethylene jacket is particularly well equipped for direct burial because it can stand up well to compressive forces. High-density polyethylene is both non-porous and non-contaminating, and provides complete protection against normal moisture and alkaline conditions.
- In many directly buried cables, an **additional moisture barrier of water-blocking gel** is applied under the jacket. If water should penetrate the jacket, it would not be able to travel under the jacket, damage would be localized and more readily repaired. Because buried cable is thermally insulated by the earth, its year-round temperature will only vary a few degrees. If a buried cable is not damaged the attenuation will be constant for the useful life of the cable.
- The **tools required** for burying cable are well-known, consisting mainly of trenchers, backhoes, and shovels.

UNDERGROUND CABLE INSTALLATION

- Because the **outer jacket is the cable's first line of defense,** any steps which can be taken to prevent damage to it will go a long way toward maintaining the internal characteristics of the cable.
- It is generally **best to bury the cable in sand or finely pulverized dirt,** without sharp stones, cinders or rubble. If the soil in the trench does not meet these requirements, tamp four to six inches of sand into the trench, lay the cable, and tamp another six to eleven inches of sand above it. A **creosoted or pressure-treated board** placed in the trench above the sand, prior to back filling, will provide some protection against subsequent damage that could be caused by digging or driving stakes.
- Lay the cable in the trench **with some slack.** A tightly stretched cable is likely to be damaged as the fill material is tamped.
- Examine the cable as it is being installed to **be sure the jacket has not been damaged** during storage, by being dragged over sharp edges on the pay-off equipment, or by other means.
- In particularly difficult installations, such as in rubble or coral, or where paving is to be installed over the cable, a **polyethylene water pipe,** which is available in long lengths and several diameters, may be buried and **used as a conduit.**
- This pipe **protects the cable** and usually makes it possible **to replace cable** which has failed without digging up the area.
- It is important that **burial is below the frost line** to avoid damage by the expansion and contraction of the earth during freezing and thawing.
- The National Electrical Code (NEC) states specific requirements for cables to be buried underground. The NEC specifies **24 inches as the minimum burial depth** for 0-600 volt nominal applications.
- If an installation must meet all NEC requirements, a **local inspector should be consulted** during the planning phase.
- Study the application, determine if any failure of the cable will result in **a hazardous condition,** and choose your cable accordingly.

ROUTING OF OUTDOOR AERIAL CIRCUITS

The code's rules or the routing of aerial coaxial and communications circuits are that such cables must:

- Be run **below power conductors** on poles.
- Remain separated from power conductors **at the attachment point** to a building.
- Have a **vertical clearance of 8 feet above roofs** (there are several exceptions).
- Coax cables run on the outside of buildings must be kept **at least 4 inches from power cables** (but not conduits).
- They must be installed so that they **do not interfere with other communications circuits;** generally, this means that they must be kept far enough away.
- Coaxial and communication cables must be kept at **least 6 feet from all lightning protection conductors,** except if such spacing is very impractical.

CIRCUIT PROTECTION

- *Protectors* are surge arresters designed for the specific requirements of communications circuits. They are required for all aerial circuits not confined with a *block.* (Block here means city block.) They must be installed on all circuits with a block that could accidentally contact power circuits over 300 volts to ground. They must also be listed for the type of installation. Other requirements are the following:

- *Metal sheaths* of any communications cables must be grounded or interrupted with an insulating joint as close as practicable to the point where they enter any building (such point of entrance being the place where the communications cable emerges through an exterior wall or concrete floor slab, or from a grounded rigid or intermediate metal conduit).

- *Grounding conductors* for communications circuits must be copper or some other corrosion-resistant material, and have insulation suitable for the area in which it is installed.

- *Communications grounding conductors* may be no smaller than No. 14.

- The grounding conductor must be run as directly as possible to the *grounding electrode,* and be protected if necessary.

- If the grounding conductor is protected by *metal raceway,* it must be bonded to the grounding conductor on both ends.

CIRCUIT PROTECTION *(cont'd)*

Grounding electrodes for communications ground may be any of the following:

1. The grounding electrode of an electrical power system.
2. A grounded interior metal piping system. (Avoid gas piping systems for obvious reasons.)
3. Metal power service raceway.
4. Power service equipment enclosures.
5. A separate grounding electrode.

If the building being served has no grounding electrode system, the following can be used as a grounding electrode:

1. Any acceptable power system grounding electrode. (See Section 250-81.)
2. A grounded metal structure.
3. A ground rod or pipe at least 5 feet long and 1/2 inch in diameter. This rod should be driven into damp (if possible) earth, and kept separate from any lightning protection system grounds or conductors.

Connections to grounding electrodes must be made with approved means.

If the power and communications systems use separate grounding electrodes, they must be bonded together with a No. 6 copper conductor. Other electrodes may be bonded also. This is not required for mobile homes.

For mobile homes, if there is no service equipment or disconnect within 30 feet of the mobile home wall, the communications circuit must have its own grounding electrode. In this case, or if the mobile home is connected with cord and plug, the communications circuit protector must be bonded to the mobile home frame or grounding terminal with a copper conductor no smaller than No. 12.

CABLE SPECIFICATIONS

The diameter of the completed cable shall be less than 6.35 mm (0.25 in)

The ultimate breaking strength of the completed cable is 90 lb minimum. Maximum *pulling* tension should not exceed 25 lb to avoid stretching.

The cable tested shall withstand a bend radius of 25.4 mm (1 in) at a temperature of −20 C without jacket or insulation cracking

The resistance of any conductor shall not exceed 28.6 ohms per 305 m (1000 ft) at or corrected to a temperature of 20 C.

The resistance unbalance between the two conductors of any pair shall not exceed 5% when measured at or corrected to a temperature of 20 C in. . . .

The mutual capacitance of any pair at 1 kHz shall not exceed 20 nF per 305 M (1000 ft).

The mutual capacitance of any pair at 1 kHz and measured at or corrected at a temperature of 20C, shall not exceed 17 nF per 305 m (1000 ft) for category 4 and category 5 cables.

The capacitance unbalance to ground at 1 kHz of any pair shall not exceed 1000 pF per 305 m (1000 ft).

CHARACTERISTIC IMPEDANCE

Characteristic impedance is another confusing telecom terms. Characteristic impedance refers to the internal signal-transmission characteristics of the cable. You cannot measure the characteristic impedance of a cable with an ohmmeter, and the number does not refer to a number of ohms per foot or per hundred feet. The really important thing with the characteristic impedance of cables is that you never switch cable with differing impedances. If you are running 100-ohm cable, you cannot insert a piece of 150-ohm cable in the network. If you were to do this, the signal would tend to reflect off of the 100-to-150-ohm junction, ruining the transmission. In short, the link will not pass signal if you mix cables of different impedance.

In general, higher impedances mean less attenuation of data signals through the cables. (Assuming that the system is designed for the purpose.)

CHARACTERISTIC IMPEDANCE *(cont'd)*

A more scientific explanation is as follows: Characteristic impedance is important because it determines the amount of power that can be transferred between devices. Whenever two electrical devices are connected, the power transferred from one to the other is maximized when their impedances are equal. (More correctly, when the output impedance of the driving device is equal to the input impedance of the driven device.) Any amount of impedance mismatch will cause some the power to be reflected back into the source device. This is why cables of differing impedance should never be directly connected, and why impedance-matching transformers are used. An unterminated piece of coax is effectively attached to an infinite impedance (being an open circuit), which is the ultimate mismatch and causes all the power arriving at the open end to be reflected. The reflections look like collisions to transmitting devices and usually bring the network to a quick halt.

There are a few cases where very short runs of cable can be mismatched, and the network not crash. A device connected to a transmission line sees only the electrical characteristics of the cable, not the device on the other end. The rule of thumb is that a conductor is a transmission line when its propagation time (the time it takes a signal to travel from one end to the other) is at least as great as the rise time of the signal being applied. So, for instance, if you assume that the rise time of a 10Mb signal is about 10 nanoseconds, and the speed of light in copper is about 1.5 ns/ft, a cable 7 feet long would act like a transmission line. That's why you can sometimes get away with using the wrong kind of cable, but only very short pieces.

SIGNAL REFLECTION

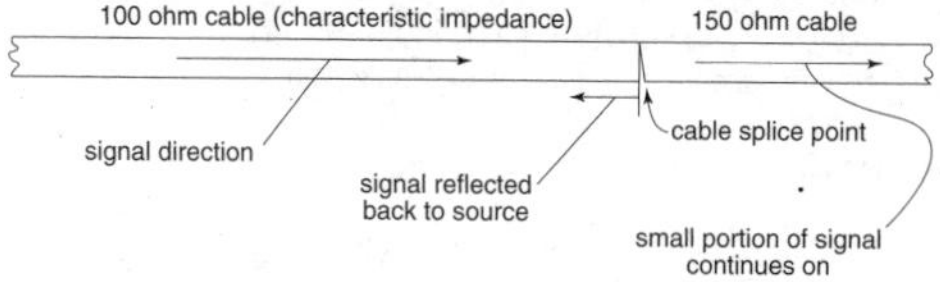

BASIC CONCEPTS OF OPTICAL FIBER

In addition to the things we have so far covered in this lesson, there are several other basic concepts that you must understand:

- **Attenuation.** Attenuation is the weakening of an optical signal as it passes through a fiber. Attenuation is signal loss. Attenuation in an optical fiber is a result of two factors, *absorption* and *scattering.*
- **Absorption** is caused by the absorption of the light and conversion to heat by molecules in the glass. Primary absorbers are residual deposits of chemicals that are used in the manufacturing process to modify the characteristics of the glass. This absorption occurs at definite *wavelengths* (the wavelength of light signifies its color and its place in the electromagnetic spectrum). This absorption is determined by the elements in the glass, and is most pronounced at wavelengths around 1000 nm (nanometers), 1400 nm and above 1600 nm.
- **Scattering.** This is the largest cause of attenuation, and occurs when light collides with individual atoms in the glass.

Fiber optic systems transmit in the "windows" created between the absorption bands at 850 nm, 1300 nm and 1550 nm wavelengths, for which lasers and detectors can be easily made.

- **Bandwidth.** Bandwidth is the range of signal frequencies or bit rate at which a fiber system can operate. In essence, it is a measure of the amount of signal that can be put through a fiber. Higher bandwidth means more data per second; lower bandwidth means less signal.
- **Dispersion.** There are two potentially confusing terms that you will come across in your readings: *Chromatic dispersion* and *modal dispersion.* In both of these terms, *dispersion* refers the spreading of light pulses, until they overlap one another, and the data signal is distorted and lost. *Chromatic* refers to color, and *modal* primarily refers to the light's path.

Thus we can state in simple terms that:

Chromatic dispersion = Signal distortion due to Color.

Modal dispersion = Signal distortion due to Path.

Note that dispersion is NOT a loss of light, it is a distortion of the signal. Thus dispersion and attenuation are two different and unrelated problems: Attenuation is a loss of light; dispersion is a distortion of the light signals.

FIBER SAFETY PRECAUTIONS

- Live optical fiber ends (*live* fibers are those with signals being sent through them) should not be inspected by technicians. **All fibers should be *dark* (no signal being transmitted) when inspected.** This must be done carefully, since the light used in the majority of optical systems is not visible to the human eye.
- If there is a risk of fibers being inspected live, especially when the system light source is a laser, all technicians working on the system should **wear protective glasses** which have infrared filtering.
- Fiber optic **work areas must be clean,** organized, well-lit, and equipped with a bottle or other suitable container for broken or stray fiber pieces.
- **No food, drink, or smoking** should be allowed in areas where fiber optic cables are spliced or terminated, or in any area where bare fibers are being handled.
- Technicians making fiber terminations or splices or working with bare fiber need double-sided tape, or some other effective means, for picking up broken or stray pieces of fiber. Their **work areas will have to be repeatedly and consistently cleared of all bare fiber pieces.** All bare fiber pieces must be disposed of so that they cannot escape and cause a hazard. (For example, bare fibers should be sealed in some type of bottle or container before being dumped into a wastebasket.)
- Finally, all technicians working on bare fiber should **thoroughly wash their hands immediately when leaving the work area,** checking their clothing, and pat themselves with clean tape to remove any stray pieces of bare fiber.

TYPICAL BANDWIDTH-DISTANCE PRODUCTS

Type of fiber	Wavelength	Bandwidth-distance product
Step-index plastic	660 nm	5 MHZ-km
multimode step index	850 nm	20 MHZ-km
multimode graded index	850 nm	600 MHZ-km
multimode graded index	1300 nm	1000-2500 MHZ-km
singlemode	1300 nm	over 300,000 MHZ-km

TYPICAL OPTICAL BUDGETS

Wavelength [nm]	Type of Source	Fiber Core Diameter [μm]	Typical Optical Power Budget, [dB]	Spectral Width [nm]
850	LED	50	12	
	LED	6.25	16	
	LED	100	21	
	LED	any		30-50
	laser	all	30	
1300	LED	50	20	
	LED	62.5	24	
	LED	100	28	
	LED	any		60-190
1300	laser [multimode]	all	50	
1300	laser [singlemode]	9	27	0.5 - 5.0

FIBER OPTIC CONNECTORS-
DATA COMMUNICATION STYLES

Style	Contact?	Keyed?	Pull Proof?	Wiggle-Proof?	Loss	Cost
ST®	Y	Y	N	N	0.3	6-10
906 SMA	N	N	Y	N	1.0	6-10
FDDI	Y	Y	Y	N	1.0	12-19
mini-BNC	N	N	N	N		10
905 SMA	N	N	Y	N	1.0	6-10
biconic	N	N	Y	N	1.0	8-25
ESCON	Y	Y	N	N	1.0	25
SC	Y	Y	Y	Y	0.3	7-14

TELEPHONE & HIGH PERFORMANCE STYLES

Style	Contact?	Keyed?	Pull Proof?	Wiggle-Proof?	Loss	Cost
ST®	Y	Y	N	N	0.3	6-10
biconic	N	N	Y	N	1.0	8-25
keyed biconic	N	Y	Y	N	1.0	8-25
FC/PC	Y	Y	Y	Y	0.3	4-15
FC	Y	Y	Y	Y	0.3	4-15
D4	Y	Y	Y	Y	0.3	12
SC	Y	Y	Y	Y	0.3	7-14

TYPICAL BANDWIDTH - DISTANCE PRODUCTS

Type of Fiber	Wavelength	Bandwidth - Distance Product
multimode step index POF	660 nm	5 MHz-km
multimode step index glass	850 nm	20 MHz-km
multimode graded index glass	850 nm	600 MHz-km
multimode graded index glass	1300 nm	1000-2500 MHz-km
singlemode glass	1310 nm	76,800-300,000 Mbps-km

OPTICAL CABLE JACKET MATERIALS & THEIR PROPERTIES

Jacket Materials	Properties
PVC	Affords normal mechanical protection. Usually specified for indoor use and general-purpose applications.
Hypalon	Has most of neoprene's properties, including ability to withstand extreme environments and flame retardancy. Has better thermal stability, and even greater oxidation and ozone resistance. Hypalon has superior resistance to radiation.
Polyethylene	Used in telephone cables. A tough, chemical- and moisture-resistant, relatively low-cost material. Since it burns, it is infrequently used in electronic applications.
Polyurethane	Has excellent abrasion resistance and low-temperature flexibility.
Thermoplastic Elastomer (TPE)	A less expensive jacketing material than neoprene or hypalon. Has many of the characteristics of rubber, along with excellent mechanical and chemical properties.
Nylon	Generally used over single conductors to improve their physical properties.

OPTICAL CABLE JACKET MATERIALS & THEIR PROPERTIES

Jacket Materials	Properties
Kynar (polyvinylidene fluoride)	A tough, abrasion- and cut- through-resistant, thermally stable and self-extinguishing material. It has low-smoke emission and is resistant to most chemicals. Its inherent stiffness limits its use as a jacket material. It has been approved for low-smoke applications.
Teflon FEP	Specified in fire alarm signal system cables. It will not emit smoke even when exposed to direct flame, is suitable for use at continuous temperatures of 200˚C and is chemically inert.
Tefzel	Like Teflon FEP, it is a fluorocarbon and has many of its properties. Rated for 150˚C, it is a tough, self-extinguishing material.
Irradiated Cross-Linked	Rated for 150˚C operation. Cross-linking changes thermoplastic polyethylene to a thermosetting material with greater resistance to environmental stress cracking, cut-through, ozone, solvents and soldering than either low- or high-density polyethylene.
Zero Halogen Thermoplastic	A thermoplastic material with excellent flame retardancy properties. Does not emit toxic fumes when it burns. Originally designed for shipboard fiber applications, it can be used for any enclosed environment.

NEC CLASSIFICATIONS FOR OPTICAL CABLES

Application	Cable marking	UL test	Jacket type
General Purpose	OFN	UL-1581	PVC or
	OFC	UL-1581	Zero Halogen
Riser	OFNR	UL-1666	PVC or
	OFCR	UL01666	Zero Halogen
Plenum	OFNP	UL-910	RVC or
	OFCP	UL-910	Fluorocarbons

Cable Marking Explanation

OFN	Optical fiber, nonconductive (all dielectric)
OFC	Optical fiber, conductive (metal strength members)
OFNR	Optical fiber, nonconductive, riser
OFCR	Optical fiber, conductive, riser
OFNP	Optical fiber, nonconductive, plenum
OFCP	Optical fiber, conductive, plenum

COMMON FIBER TYPES

Item Number	Diameter Core	Diameter Cladding	Index Profile	Primary Buffer Diameter	Attenuation (dB/km)	Bandwidth (MHz/km)	Numerical Aperture
1)	50µm	125µm	Graded	250µm	•3@0.85µm •1@1.3µm	200-800@0.85µm 200-800@1.3µm	0.20
2)	50µm	125µm	Graded	500µm	•4@0.85µm •2@1.3µm	200-800@0.85µm 200-800@1.3µm	0.20
3)	62.5µm	125µm	Graded	250µm	•3.5@0.85µm •1.5@1.3µm	100-300@0.85µm 100-800@1.3µm	0.275
4)	62.5µm	125µm	Graded	500µm	•3.5@0.85µm •1.5@1.3µm	100-300@0.85µm 100-800@1.3µm	0.275
5)	85µm	125µm	Graded	500µm	•4@0.85µm •2@1.3µm	100-200@0.85µm 200-400@1.3µm	0.26
6)	100µm	140µm	Graded	500µm	•5@0.85µm •3@1.3µm	100-300@0.85µm 100-500@1.3µm	0.30
7)	100µm	140µm	Step	500µm	•10@0.85µm	20@0.85µm	0.24
8)	100µm	140µm	Graded	160µm	•6@0.85µm	100@0.85µm	0.30
9)	200µm	230µm	Step(HCS)	500µm	•8@0.85µm	17@0.85µm	0.37
10)	200µm	240µm	Step	500µm	•10@0.85µm	20@0.85µm	0.24
11)	200µm	380µm	Step(PCS)	600µm	•10@0.85µm	8@0.85µm	0.40

Note items 7 and 9 are radiation-hard fibers. Item 8 is a high-temperature fiber.

COMPARISON OF LASER & LED LIGHT SOURCES

Characteristic	LED	Laser
Output power	Lower	Higher
Speed	Slower	Faster
Output pattern (NA)	Higher	Lower
Spectral width	Wider	Narrower
Single-mode compatibility	No	Yes
Ease of Use	Easier	Harder
Cost	Lower	Higher

COMPARISON OF BUFFER TYPES

Cable	Cable Structure	
Parameter	Loose Tube	Tight Buffer
Bend Radius	Larger	Smaller
Diameter	Larger	Smaller
Tensile Strength, Installation	Higher	Lower
Impact Resistance	Lower	Higher
Crush Resistance	Lower	Higher
Attenuation Change At Low Temperatures	Lower	Higher

INDICES OF REFRACTION

Material	Index	Light Velocity(km/s)
Vacuum	1.0	300,000
Air	1.0003	300,000
Water	1.33	225,000
Fused quartz	1.46	205,000
Glass	1.5	200,000
Diamond	2.0	150,000
Silicon	3.4	88,000
Gallium arsenide	3.6	83,000

TRANSMISSION RATES - DIGITAL TELEPHONE

Medium	Bit Rate (Mbps)	Voice Channels	Repeater Spacing (km)
Coaxial	1.5	24	1-2
	3.1	48	
	6.3	96	
	45	672	
	90	1,344	
Fiber	45	672	6-15 (multimode)
	90	1,344	30-40+ (single mode)
	180	2,688	
	405 to 435	6,048	
	565	8,064	
	1,700	24,192	

MISMATCHED FIBER CONNECTION LOSSES (Excess loss in dB)

	Transmitting Fiber		
Receiving Fiber	62.5/125	85-125	100/140
50-125	0.9-1.6	3.0-4.6	4.7-9
62.5/125	—	0.9	2.1-4.1
85/125	—	—	0.9-1.4

POWER LEVELS OF FIBER OPTIC COMMUNICATION SYSTEMS

Network Type	Wavelength (nm)	Power Range (dBm)	Power Range (W)
Telecom	1300, 1550	+3 to -45	50nW to 2mW
Datacom	665, 790, 850, 1300	-10 to -30	1 100uW
CATV	1300, 1550	+10 to -6	250uW to 10mW

DETECTORS USED IN FIBER OPTIC POWER METERS

Detector Type	Wavelength Range (nm)	Power Range (dBm)	Comments
Silicon	400-1100	+10 to -70	
Germanium	800-1600	+10 to -60	-70 with small area detectors, +30 with attenuator windows
InGaAs	800-1600	+10 to -70	Small area detectors may overload at high power (>.0 dBm)

FIBER OPTIC TESTING REQUIREMENTS

Test Parameter	Instrument
Optical power (source output, receiver signal level)	Fiber optic power meter
Attenuation or loss of fibers, cables, and connectors	FO power meter and source, test kit or OLTS (optical loss test set)
Source wavelength*	FO spectrum analyzer
Backscatter (loss, length, fault location)	Optical time domain reflectometer (OTDR)
Fault location	OTDR, visual cable fault locator
Bandwidth/dispersion* (modal and chromatic)	Bandwidth tester or simulation software

* Rarely tested in the field

TYPICAL CABLE SYSTEM FAULTS

Fault	Cause	Equipment	Remedy
Bad connector	Dirt or damage	Microscope	Cleaning/ polishing retermination
Bad pigtail	Pigtail kinked	Visual fault locator	Straighten kink
Localized cable attenuation	Kinked cable	OTDR	Straighten kink
Distributed increase in cable attenuation	Defective cable or installation specifications exceeded	OTDR	Reduce stress/ replace
Lossy splice	Increase in splice Loss due to fiber stress in closure	OTDR Visual fault locator	Open and redress
Fiber break	Cable damage	OTDR Visual fault locator	Repair/replace

MOST COMMON CAUSES OF FAILURES IN FIBER OPTIC LANS

1 Broken fibers at connector joints
2 Broken fibers at patch panels
3 Cables damaged at patch panels
4 Fibers broken at patch panels
5 Cables cut in ceilings and walls
6 Cables cut through outside construction
7 Contaminated connections
8 Broken jumpers
9 Too much loss
10 Too little loss (overdriving the receiver)
11 Improper cable rolls
12 Miskeyed connectors
13 Transmission equipment failure
14 Power failure

FIBER OPTIC LOSS BUDGET ANALYSIS

Loss budget analysis is the calculation of a fiber optic system's operating characteristics. This encompasses items such as routing, electronics, wavelengths, fiber type, and circuit length. Attenuation and bandwidth are the key parameters for budget loss analysis.

Prior to implementing or designing a fiber optic circuit, a loss budget analysis is recommended to make certain the system will work over the proposed link. Both the passive and active components of the circuit have to be included in the budget loss calculation. Passive loss is made up of fiber loss, connector loss, and splice loss. Don't forget any couplers or splitters in the link. Active components are system gain, wavelength, transmitter power, receiver sensitivity, and dynamic range. Prior to system turn up, test the circuit with a source and FO power meter to ensure that it within the loss budget.

The idea of a loss budget is to insure the network equipment will work over the installed fiber optic link. It is normal to be conservative. Don't use the best possible specs for fiber attenuation or connector loss.

The best way to illustrate calculating a loss budget is to show how it's done for a 2 km multimode link with 4 connections (2 connectors at each end and 2 connections at patch panels in the link) and one splice in the middle.

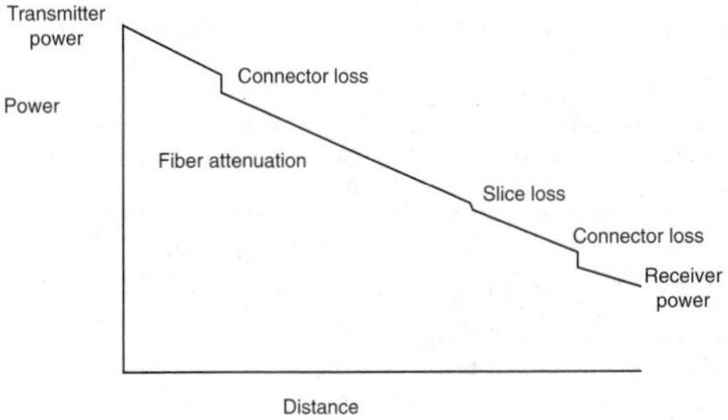

CABLE PLANT PASSIVE COMPONENT LOSS

Step 1. Fiber loss at the operating wavelength

Fiber Type	Multimode	Singlemode
Wavelength (nm)	850	1300
Fiber Atten. dB/km	3 (3.5)	1 (1.5)

For a 2km length of multimode fiber operating at 850 nm, the attenuation would be:

2 km x 3 dB/km = 6 dB attenuation

(All specs in brackets are maximum values per EIA/TIA 568 standard. For singlemode fiber, a higher loss is allowed for premises applications.)

Step 2. Connector Loss:

Multimode connectors will have losses of 0.2-0.5 dB Singlemode connectors, which are factory made and fusion spliced on will have losses of 0.1-0.2 dB. Field terminated singlemode connectors may have losses as high as 0.5-1.0 dB.

Typical Connector Loss	0.5 dB
Total # of Connectors	4
Total Connector Loss	2.0 dB

(All connectors are allowed 0.75 max per EIA/TIA 568 standard)

Step 3. Splice Loss:

Multimode splices are usually made with mechanical splices, although some fusion splicing is used. The larger core and multiple layers make fusion splicing about the same loss as mechanical splicing, but fusion is more reliable in adverse environments. Figure 0.1-0.5 dB for multimode splices, 0.2 being a good average for an experienced installer. Fusion splicing of singlemode fiber will typically have less than 0.05 dB

Typical Splice Loss	0.2 dB
Total # Splices	1
Total Splice Loss	0.2 dB

(All splices are allowed 0.3 max per EIA/TIA 568 standard)

CABLE PLANT PASSIVE COMPONENT LOSS *(cont'd)*

Step 4. Total Passive System Attenuation: Add the fiber loss, connector and splice losses to get the link loss.

Wavelength (NM)	850	1300
Total fiber Loss (dB)	6	2.0
Total Connector Loss (dB)	2.0	2.0
Total Splice Loss (dB)	0.2	0.2
Other (dB)	0	0
Total Link Loss (dB)	8.2	4.2

Note: Remember these should be the criteria for testing. Allow +/- 0.2 -0.5 dB for measurement uncertainty and that becomes your pass/fail criterion.

Equipment Link Loss Budget Calculation: Link loss budget for network hardware depends on the dynamic range, the difference between the sensitivity of the receiver and the output of the source into the fiber. You need some margin for system degradation over time or environment, so subtract that margin (as much as 3dB) to get the loss budget for the link.

Step 5. From Manufacturer's Specification for Active Components

Operating Wavelength (nm)	1300
Fiber Type	MM
Receiver Sens. (dBm@BER)	-31
Average Transmitter Output (dBm)	-16
Dynamic Range (dB)	15
Recommended Excess Margin (dB)	3
Max Cable Plant Loss Budget (dB)	12

Step 6. Loss Margin Calculation

Dynamic Range (dB)	15
Cable Plant Link Loss (dB)	4.2
Link Loss Margin (dB)	10.8

As a general rule, the Link Loss Margin should be greater than approximately 3 dB to allow for link degradation over time. LEDs in the transmitter may age and lose power, connectors or splices may degrade or connectors may get dirty if opened for rerouting or testing. If cables are accidentally cut, excess margin will be needed to accommodate splices for restoration.

SPECIFICATIONS FOR FIBER OPTIC NETWORKS

Application	Wavelength	Max Distance (m) for fiber type			Link Margin (dB) for fiber type	
		62.5	50	SM	62.5	50
10Base-F	850	2000	2000	NS	12.5	7.8
FOIRL	850	2000	NS	NS	8	NS
Token Ring 4/16	850	2000	2000	NS	13	8.3
Demand Priority (100VG-AnyLAN)	850	500	500	NS	7.5	2.8
Demand Priority (100VG-AnyLAN)	1300	2000	2000	NS	7.0	2.3
100Base-FX (Fast Ethernet)	1300	2000	2000	NS	11	6.3
100Base-SX	850	300	300	NS	4.0	4.0
FDDI	1300	2000	2000	40000	11.0	6.3
FDDI (low cost)	1300	500	500	NA	7.0	2.3

NA = Not Applicable NS = Not Specified

SPECIFICATIONS FOR FIBER OPTIC NETWORKS *(cont'd)*

Application	Wavelength	Max Distance (m) for fiber type			Link Margin (dB) for fiber type	
		62.5	50	SM	62.5	50
ATM 52	1300	3000	3000	15000	10	5.3
ATM 155	1300	2000	2000	15000	10	5.3
ATM 155	850 (laser)	1000	1000	NA	7.2	7.2
ATM 622	1300	500	500	15000	6.0	1.3
ATM 622	850 (laser)	300	300	NA	4.0	4.0
Fibre Channel 266	1300	1500	1500	10000	6.0	5.5
Fibre Channel 266	850 (laser)	700	2000	NA	12.0	12.0
Fibre Channel 1062	850 (laser)	300	500	NA	4.0	4.0
Fibre Channel 1062	1300	NA	NA	10000	NA	NA
1000Base-SX	850 (laser)	220	550	NA	3.2	3.9
1000Base-LX	1300	550	550	5000	4.0	3.5
ESCON	1300	3000	NS	20000	11	NS

NA = Not Applicable NS = Not Specified

FIBER OPTIC LABOR UNITS

Labor Item	Labor Units (Hours) Normal	Difficult
Optical fiber cables, per foot:		
1-4 fibers, in conduit	0.016	0.02
1-4 fibers, accessible locations	0.014	0.018
12-24 fibers, in conduit	0.02	0.025
12-24 fibers, accessible locations	0.018	0.023
48 fibers, in conduit	0.03	0.038
48 fibers, accessible locations	0.025	0.031
72 fibers, in conduit	0.04	0.05
72 fibers, accessible locations	0.032	0.04
144 fibers, in conduit	0.05	0.065
144 fibers, accessible locations	0.04	0.05
Hybrid cables:		
1-4 fibers, in conduit	0.02	0.025
1-4 fibers, accessible locations	0.017	0.021
12-24 fibers, in conduit	0.024	0.03
12-24 fibers, accessible locations	0.022	0.028
Testing, per fiber	0.10	0.20
Splices, including prep and failures, trained workers:		
fusion	0.20	0.30
mechanical	0.30	0.40
array splice, 12 fibers	0.80	1.20
Coupler (connector-connector)	0.15	0.25
Terminations, including prep and failures, trained workers:		
polishing required	0.35	0.55
no-polish connectors	0.25	0.40
FDDI dual connector, including terminations	0.70	0.95
Miscellaneous:		
cross-connect box, 144 fibers, not including splices	3.00	4.00
splice cabinet	2.00	2.50
splice case	1.80	2.25
breakout kit, 6 fibers	1.00	1.40
tie-wraps	0.01	0.02
wire markers	0.01	0.01

FIBER OPTIC SAFETY RULES

1. Keep all food and beverages out of the work area. If fiber particles are ingested, they can cause internal hemorrhaging.
2. Wear disposable aprons to minimize fiber particles on your clothing. Fiber particles on your clothing can later get into food, drinks, and/or be ingested by other means.
3. Always wear protective gloves and safety glasses with side shields. Treat fiber optic splinters the same as you would glass splinters.
4. Never look directly into the end of fiber cables until you are positive that there is no light source at the other end. Use a fiber optic power meter to make certain the fiber is dark. When using an optical tracer or continuity checker, look at the fiber from an angle at least six inches away from your eye to determine if the visible light is present.
5. Only work in well ventilated areas.
6. Contact wearers must not handle their lenses until they have thoroughly washed their hands.
7. Do not touch your eyes while working with fiber optic systems until your hands have been thoroughly washed.

COMMON FIBER OPTIC CONNECTORS

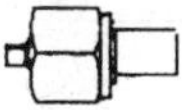

SFR Type

SMA905

SMA906

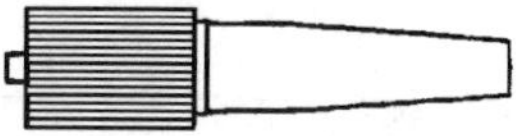

D4 CERAMIC CONNECTOR

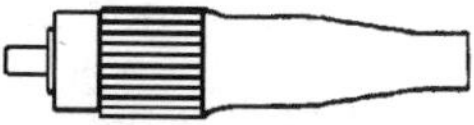

***FC** Cylindrical with metal coupling*

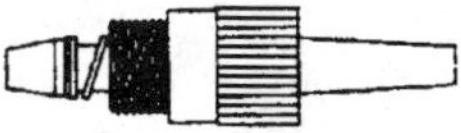

BICONIC

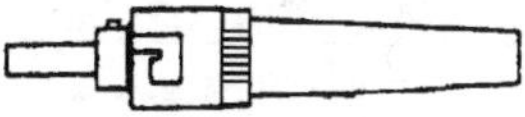

***ST** Cylindrical with twist lock coupling*

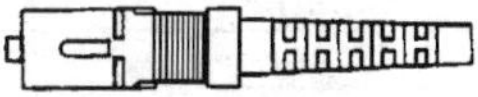

***SC** Square, keyed connector*

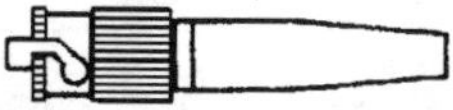

***MINI BNC** Cylindrical with twist lock coupling*

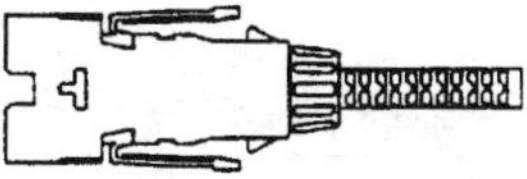

FDDI
Duplex connector, with fixed shroud, keyed.

ESCON
Duplex connector, with retractable shroud.

OPTICAL CABLE CRUSH STRENGTHS

Characteristic	Type of Cable	Pounds/Inch
Long-term crush load	>6 fibers/cable	57-400
	1-2 fiber cables	314-400
	Armored cables	450
Short-term crush load	>6 fibers/cable	343-900
	1-2 fiber cables	300-800
	Armored cables	600

MAXIMUM VERTICAL RISE DISTANCES

Application	Feet
1 fiber in raceway or tray	90
2 fiber in duct or conduit	50-90
Multifiber (6-12) cables	50-375
Heavy duty cables	1000-1640

MAXIMUM RECOMMENDED INSTALLATION LOADS

Application	Pounds Force
1 fiber in raceway or tray	67
1 fiber in duct or conduit	125
2 fiber in duct or conduit	200
Multifiber (6-12) cables	250-500
Direct burial cables	600-800
Lashed aerial cables	>300
Self-support aerial cables	>600

TYPICAL TEMPERATURE RANGES OF OPERATION

Optical Cable Application	Temp Range, degrees C.
indoor	−10 to +60, −10 to +50
outdoor	−20 to +60, −40 to +50, −40 to +70
military	−55 to +85
aircraft	−62 to +125

CABLE SELECTION CRITERIA

1. Current and future bandwidth requirements
2. Acceptable attenuation rate
3. Length of cable
4. Cost of installation
5. Mechanical requirements (ruggedness, flexibility, flame retardance, low smoke, cut-through resistance)
6. UL/NEC requirements
7. Signal source (coupling efficiency, power output, receiver sensitivity)
8. Connectors and terminations
9. Cable dimension requirements
10. Physical environment (temperature, moisture, location)
11. Compatibility with any existing systems

FIBER OPTIC DATA NETWORK STANDARDS

Network	**IEEE802.3 FOIRL**	**IEEE802.3 10baseF**	**IEEE802.5 Token Ring**	**ANSI X3T9.5 FDDI**	**ESCON IBM**
Bitrate (MB/s)	10	10	4/16	100	200
Architecture	Link	Star	Ring	Ring	Branch
Fiber type	MM, 62.5	MM, 62.5	MM, 62.5	MM/SM	MM/SM
Link length (km)	2	—	—	2/60	3/20
Wavelength (nm)	850	850	850	1300	1300
Margin (dB, MM/SM)	8	—	12	11/27	8*(11)/16
Fiber FW (mHz-km)	150	150	150	500	500
Connector	SMA	ST	FDDI	FDDI	ESCON

FIBER TYPES AND SPECIFICATIONS

Fiber Type	**Core/Cladding Diameter (m)**	**Attenuation 850 nm**	**Coefficient 1300 nm**	**(dBkm) 1550 nm**	**Bandwidth (MHz-km)**
Multimode/Plastic	1mm	(1 dB/m	@665 nm)		Low
Multimode/Step Index	200/240	6			50
Multimode/Graded Index	50/125	3	1		600
	62.5/125	3	1		500
	85/125	3	1		500
	100/140	3	1		300
Single mode	8-9/125		0.5	0.3	high

GENERAL DESIGN FACTORS FOR OPTICAL CABLING SYSTEMS

The general factors to be considered when choosing an optic fiber cable plant are as follows:

1. Current and future bandwidth requirements.
2. Acceptable attenuation rate.
3. Length of cable.
4. Cost of installation.
5. Mechanical requirements (ruggedness, flexibility, flame retardance, low smoke, cut-through resistance).
6. Code and safety requirements.
7. Signal source (coupling efficiency, power output, receiver sensitivity),
8. Connectors and terminations.
9. Cable dimension requirements.
10. Physical environment (temperature, moisture, location).
11. Compatibility with any existing systems.

DESIGN CHOICES

The following pages illustrate design choices (short cuts) normally used for optical cabling systems. These are generally the most applicable and cost-effective design choices in an open marketplace. Note, however, that these are general recommendations, and that situations may dictate otherwise.

FIBER CHOICE

- **Multimode—Use 62.5/125 fiber.** 62.5/125 is the de facto standard. It offers higher launched power than the 50/125 fiber at a modest price increase. In addition, most standards for data communication are focused the 62.5/125 fiber. There is no indication that this will change any time soon.
- **Singlemode—Use 1300 nm singlemode fiber.** Systems designed to operate at this wavelength have lower cost than do 1550 nm systems. Do not choose fiber designed for both 1300 and 1550 nm unless you expect to use wavelength division multiplexing or optical amplifiers in the future.

CABLE DESIGN

- **Indoor—For short distances [<1200-1335 feet], use break-out cables. For longer distances, use premise (distribution) cables.** If your environment is especially rugged, use break-out design rather than premise cable. The price premium is insurance against future maintenance cost.
- **Use all-dielectric design.** These cables are not subject to grounding requirements. **If plenum cables required, look for plenum-rated PVC products.** Teflon plenum cables are not only more expensive, but they are often hard to find or will be available only after a long delay.
- **Outdoor cables—Use water-blocked and gel-filled, loose tube designs.** Consider ribbon designs (also water-blocked and gel-filled) if the cable will have 36 or more fibers. Ribbon-based cables are cheaper *per fiber* than other types of cables. **If mid-span access is important, use the stranded loose-tube design. Use all-dielectric design.** For the reason specified earlier—no grounding requirements.
- **Indoor/Outdoor Cables—If cable must be installed both indoors and outdoors, use indoor/outdoor rated cable.** In so doing, you can eliminate a splice or connector pair. This design has an easily removable outdoor jacket over an inner structure which meets NEC requirements.

FIBER PERFORMANCE

- **Multimode-Choose dual wavelength specifications.**

Wavelength	850/1300 nm
Attenuation rate	3.75/1.0 dB/km
Bandwidth-distance product	160/500 MHZ-km
NA (numerical aperture)	.275 nominal

- **Singlemode-Choose single wavelength specifications.**

Wavelength	1300 nm
Attenuation rate	0.5 dB/km
Dispersion	3.5 ps/km/nm @ 1310 nm

CABLE PERFORMANCE

- **Indoor Maximum recommended installation load 360-500 pounds**
  —Temperature operating range −10 to 60 degrees C.

- **Outdoor-Maximum recommended installation load 600 pounds**
  —Temperature operating range −40 to 60 degrees C.
  —If rodent resistance is required, use armored cable or install cable in inner duct
  —Strength members—Use epoxy-fiberglass or flexible fiberglass
  —Jacket material—Black polyethylene

INDUSTRY OPTICAL STANDARDS

Aside from NEC fire-rating classifications such as general-duty, riser-rated cable, and plenum-rated cable, the physical construction of optical cables is not governed by any agency. It is up to the designer of the system to make sure that the cable selected will meet the application requirements. Five basic cable types have however emerged as de facto standards for a variety of applications:

- **Simplex and zip cord:** One or two fibers, tight-buffered, Kevlar reinforced and jacketed. Used mostly for patch cord and backplane applications.
- **Distribution cables:** (Also known as tightpack cables.) Up to several tight-buffered fibers bundled under the same jacket with Kevlar reinforcement. Used for short, dry conduit runs, riser and plenum applications. These cables are small in size, but because their fibers are not individually reinforced, these cables need to be terminated inside a patch panel or junction box.
- **Breakout cables:** These cables are made of several simplex units, cabled together. This is a strong, rugged design, and is larger and more expensive than the tightpack cables. It is suitable for conduit runs, riser and plenum applications. Because each fiber is individually reinforced, this design allows for a strong termination to connectors and can be brought directly to a computer backplane.
- **Loose tube cables:** These are composed of several fibers cabled together, providing a small, high fiber count cable. This type of cable is ideal for outside plant trunking applications. Depending on the actual construction, it can be used in conduits, strung overhead or buried directly into the ground.
- **Hybrid or composite cables:** There is a lot of confusion over these terms, especially since the 1993 United States National Electrical Code switched their terminology from "hybrid" to "composite".

  —Under the new terminology, a *composite* cable is one that contains a number of copper conductors properly jacketed and sheathed depending on the application, in the same cable assembly as the optical fibers.

  —This situation is made all the more confusing since there is another type of cable that was formerly called composite. This type of cable contains only optical fibers, but have two different types of fibers: Multi-mode and single-mode.

  —Remember that there is confusion over these terms; with some people using them interchangeably. At this point the proper terminology is the following:

  A *composite* cable is a fiber/copper cable.

  A *hybrid* cable is a fiber/fiber cable.

INDUSTRY OPTICAL WORK STANDARDS

- **Standard wavelengths** for optical fiber systems are 850nm (nanometers), 1300nm, and 1550 nm. (For those who are unfamiliar, the various wavelengths are simply slightly different colors of light. Rather than calling them "deep red, slightly deep red, etc., we state the wavelength of each color, which we measure in nanometers.)

- Most fiber optic telecom systems in the US are based on **single mode fiber** and 1300 nm laser light with a very pure color. They are, however, rather expensive.)

- Most fiber data networks are **multimode systems,** powered by 850 nm LED light sources. (Light Emitting Diodes are not as powerful as lasers, and can not put out a pure color. They are none the less widely used, being very affordable.)

- Type ST connectors are usually used, but are being **slowly replaced by SC connectors.**

- Although there are several types of multimode fibers that have been used—50/125, 62.5/125, 85/125 and 100/140 (core/clad measurements, in microns), **62.5/125 fiber is now almost exclusively used.**

BLOWN-IN FIBER

When working with multiple innerducts inside a conduit, one of the quickest methods of installing optical fiber is to blow it in. When done properly, this method works very well. The process work as follows:

- A compressor with about **140 PSI** supplies air to the jetting machine. Air is supplied to one innerduct to clean out any possible debris.
- **Small foam sponges** are shot through each innerduct to further clean the system.
- A few ounces of **lubricant** are added to the already pre-lubricated PVC innerducts.
- **A tarp is laid out** to keep the cable clean before innerduct entry.
- A **small hollow "bullet"** is attached to the end of the cable for alignment.
- The fiber cable is blown into the duct at **up to 200 feet per minute** at a pressure of **110 PSI to 120 PSI.** (Speeds of 300+ feet are obtainable using pneumatic units.)
-

CABLE ASSEMBLY SPECIFICATIONS

Connector	SC	SC Angled	ST*
Attenuation 1300 (dB)			
Single-mode	Mean 0.25, sigma 0.1	Mean 0.25, sigma 0.1	Mean 0.25, sigma 0.15
Multimode	Mean 0.15, sigma 0.05	—	Mean 0.15, sigma 0.05
Reflection (dB)			
Single-mode	≤−50, Mean −58	≤−70, Mean −80	≤−50, Mean −58
Multimode	≤−25, Mean −27	—	≤−25, Mean −27
Connection durability (dB)			
Single-mode	<0.2 change	<0.1 change	<0.1 change
Multimode	<0.2 change	—	<0.2 change
Number of matings			
Single-mode	500	500	500
Multimode	500	—	500
Operational temperature			
Connector only (cable dependent)			
Single-mode	−40° to 85°C	−40° to 85°C	−40° to 85°C
Multimode	−40° to 85°C	—	−40° to 85°C
Storage temperature			
Single-mode	−40° to 85°C	−40° to 85°C	−40° to 85°C
Multimode	−40° to 85°C	—	−40° to 85°C
Material			
Connector plug	—	—	—
Connector housing	Engineering Thermoplastic	Engineering Thermoplastic	Nickel plated zinc
Connector ferrule	Zirconia ceramic	8° angle Zirconia ceramic	Zirconia ceramic
Alignment sleeve			
Single-mode	Zirconia ceramic	Zirconia ceramic	Zirconia ceramic
Multimode	Zirconia ceramic	—	Metal
Boot	Polyester	Polyester	Estane®
Flame retardant	UL-94 V-O	UL-94 V-O	UL-94 V-O

CABLE ASSEMBLY SPECIFICATIONS *(cont'd)*

	ST* Push-Pull	FC/PC	FC Angled	Biconic	FDDI
	Mean 0.25, sigma 0.15	Mean 0.25, sigma 0.15	Mean 0.20 sigma 0.1	Mean 0.35, sigma 0.15	—
	Mean 0.15, sigma 0.05	Mean 0.15, sigma 0.05	—	Mean 0-.5, sigma 0.15	0.6 dB max., 0.25 dB typ.
	≤−50, Mean −58	≤−50, Mean −58	≤−70, Mean −80	−30 typical	—
	≤−25, Mean −27	≤−25, Mean −27	—	−14 typical	—
	<0.2 change	<0.2 change	<0.1 change	<0.2 change	—
	<0.2 change	<0.2 change	—	<0.2 change	<0.2 change
	500	500	500	500	—
	500	500	—	1000	500
	−10° to 40°C	−40° to 85°C	−40° to 85°C	−40° to 80°C	—
	−10° to 40°C	−40° to 85°C	—	−40° to 80°C	−40° to 80°C
	−40° to 85°C	−40° to 85°C	−40° to 85°C	−40° to 85°C	—
	−40° to 85°C	−40° to 85°C	—	−40° to 85°C	−40° to 85°C
Material					
	—	—	—	70% Silica filled epoxy	—
	Engineering Thermoplastic	Engineering Thermoplastic	Engineering Thermoplastic	Engineering Thermoplastic	—
	Zirconia ceramic	Zirconia ceramic	8° angle Zirconia ceramic	—	Zirconia ceramic
	Zirconia ceramic	Zirconia ceramic	Zirconia ceramic	70% Silica filled epoxy	—
	Metal	Zirconia ceramic	—	70% Silica filled epoxy	Metal
	Santoprene	Polyester	Polyester	Estane®	Estane®
	UL-94 V-O	UL-94 V-O	UL-94 V-O	UL-94 V-O	UL-94 V-O

CHAPTER 5
TESTING

EIA/TIA 568 TESTING DISTANCE SPECIFICATION

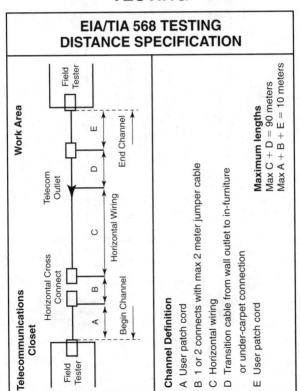

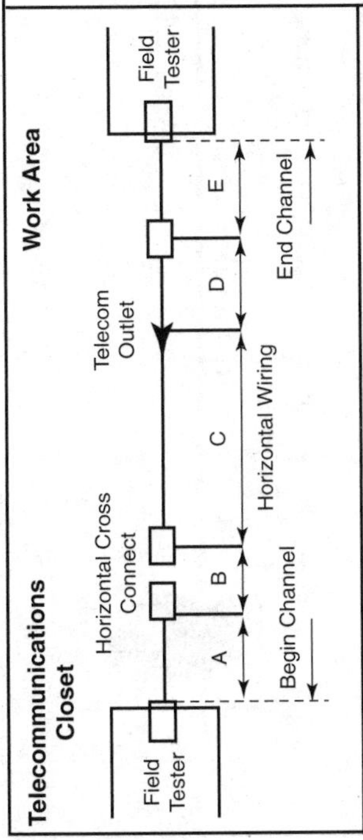

Channel Definition

A User patch cord
B 1 or 2 connects with max 2 meter jumper cable
C Horizontal wiring
D Transition cable from wall outlet to in-furniture or under-carpet connection
E User patch cord

Maximum lengths

Max C + D = 90 meters
Max A + B + E = 10 meters

EIA/TIA 568 LINK DEFINITION

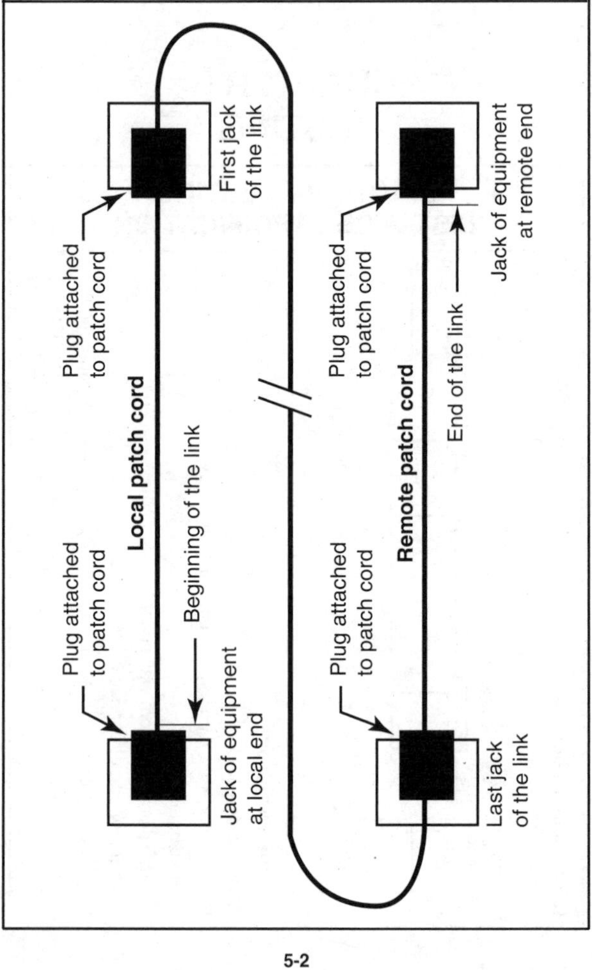

EIA/TIA LINK CONFIGURATION

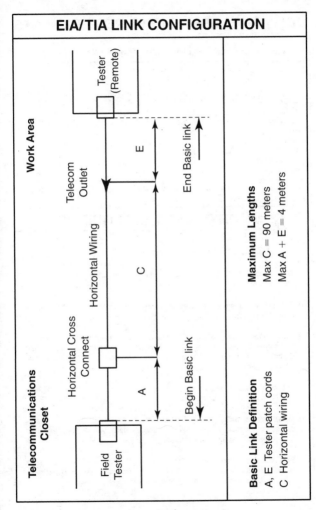

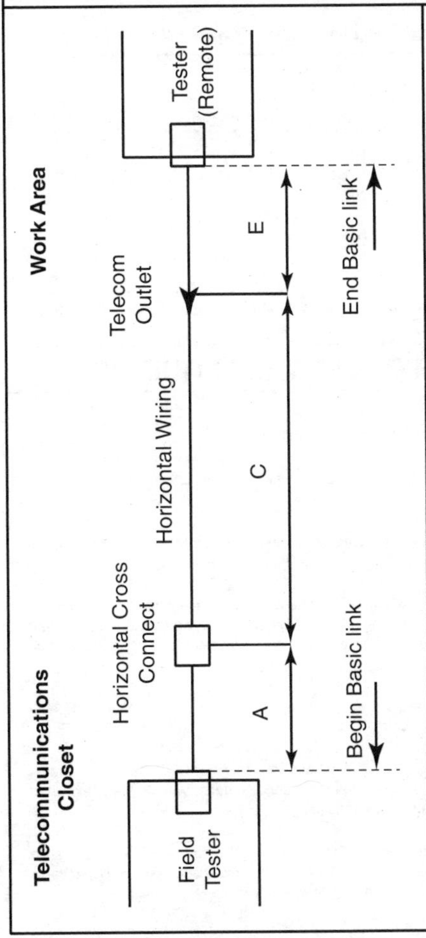

Basic Link Definition

A, E Tester patch cords

C Horizontal wiring

Maximum Lengths

Max C = 90 meters

Max A + E = 4 meters

NETWORK CABLING TEST REQUIREMENTS

Cable/Network	Cat 1,2	Cat 3	Cat 5
Analog phone	W	W	W
Digital Phone/PBX	1-NR, Cat2-W	W	W
Ethernet	NR	L, W, X, A	L, W, X. A
4 MB Token Ring	NR	L, W, X, A	L, W, X, A
16 MB Token Ring	NR	L, W, X, A	L, W, X, A
100Base-T4	NR	NR	L, W, X, A, PD, S
100Base-Tx	NR	NR	L, W, X, A, PD
100VG AnyLAN	NR	L, W, X, A, S	L, W, X, A, S
TP-PMD/FDDI	NR	NR	L, W, X, A
155 MB/s ATM	NR	NR	L, W, X, A
GB Ethernet	NR	NR	Undefined

Tests: L = length, W = wire map of connections, X = NEXT (crosstalk), A = attenuation, S = delay skew, PD = propagation delay, NR = not recommended

COMMON CABLE TEST EQUIPMENT

- **DVM *(Digital Volt Meter).*** Measures volts.
- **DMM *(Digital Multi Meter).*** Measures volts, ohm, capacitance, and some measure frequency.
- **TDR *(Time Domain Reflectometer).*** Measures cable lengths, locates impedance mismatches.
- **Tone Generator and Inductive Amplifier.** Used to trace cable pairs, follow cables hidden in walls or ceiling. The tone generator will typically put a 2 kHz audio tone on the cable under test, the inductive amp detects and plays this through a built-in speaker.
- **Wiremap Tester.** Checks a cable for open or short circuits, reversed pairs, crossed pairs and split pairs.
- **Noise testers, 10Base-T.** The standard sets limits for how often noise events can occur, and their size, in several frequency ranges. Various handheld cable testers are able to perform these tests.
- **Butt sets.** A telephone handset that when placed in series with a battery (such as the one in a tone generator), allows voice communication over a copper cable pair. Can be used for temporary phone service in a wiring closet.

ATTENUATION-TO-CROSSTALK RATIO (ACR)

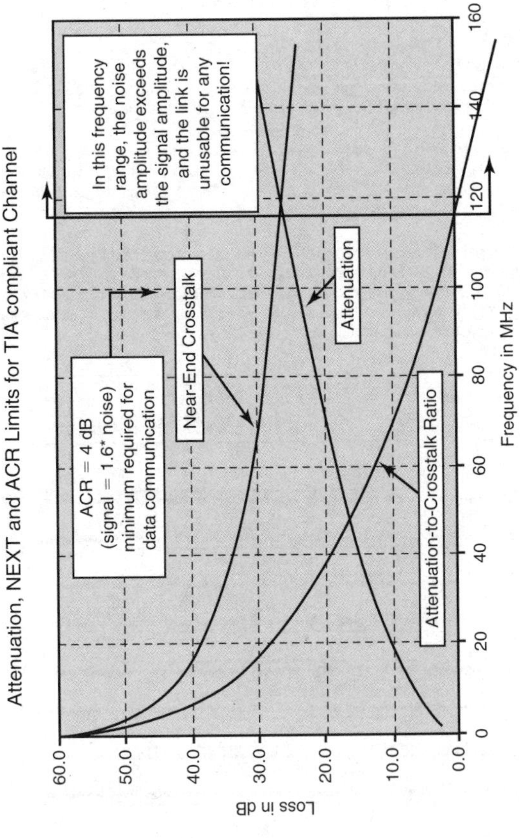

Attenuation, NEXT and ACR Limits for TIA compliant Channel
In this frequency range, the noise amplitude exceeds the signal amplitude, and the link is unusable for any communication!
ACR = 4 dB (signal = 1.6* noise) minimum required for data communication
Near-End Crosstalk
Attenuation
Attenuation-to-Crosstalk Ratio
60.0
50.0
40.0
30.0
20.0
10.0
0.0
Loss in dB
0
20
40
60
80
100
120
140
160
Frequency in MHz

WIREMAPPING

Wiremap tests will check all lines in the cable for all of the following errors:

- **Open:** Lack of continuity between pins at both ends of the cable.
- **Short:** Two or more lines short-circuited together.
- **Crossed pair:** A pair is connected to different pins at each end (example: pair 1 is connected to pins 4 & 5 at one end, and pins 1 & 2 at the other).
- **Reversed pair:** The two lines in a pair are connected to opposite pins at each end of the cable. For example: the line on pin 1 is connected to pin 2 at the other end, the line on pin 2 is connected to line 1. This is also called a *polarity reversal or tip-and-ring reversal.*
- **Split pair:** One line from each of two pairs is connected as if it were a pair. For example, the Blue and White-Orange lines are connected to pins 4 & 5, White-Blue and Orange pins to 3 & 6. The result is excessive Near End Crosstalk (NEXT), which wastes 10Base-T bandwidth and usually prevents 16 Mb/s token-ring from working at all.

EIA/TIA 568A MODULAR PIN CONNECTIONS

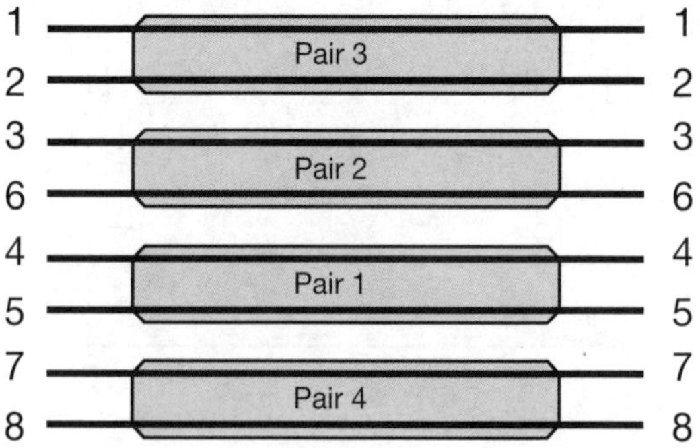

MODULAR PIN CONNECTIONS- REVERSED PAIR

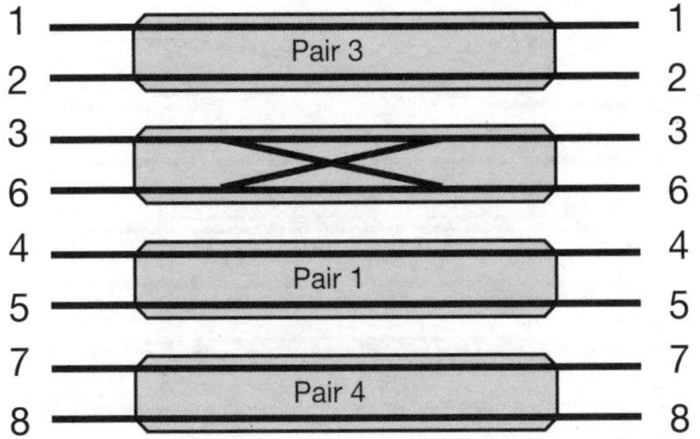

MODULAR PIN CONNECTIONS- SHORTS AND OPENS

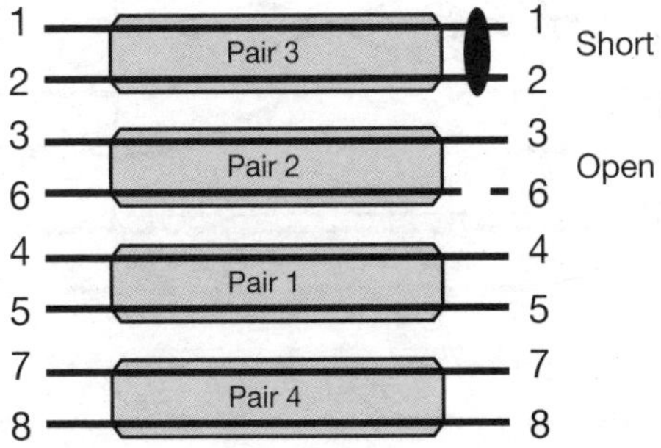

MODULAR PIN CONNECTIONS-SPLIT PAIRS

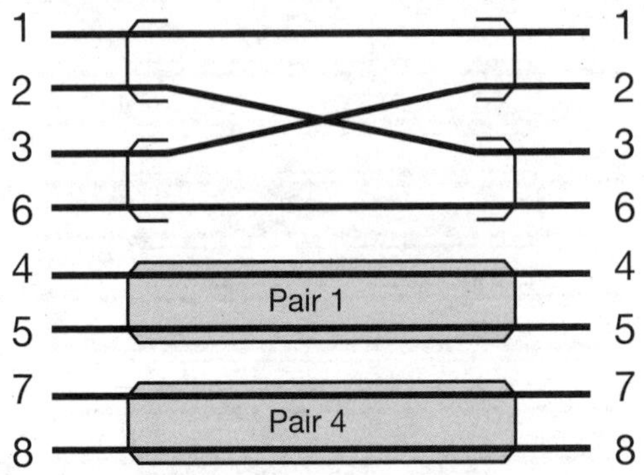

MODULAR PIN CONNECTIONS-TRANSPOSED OR CROSSED PAIRS

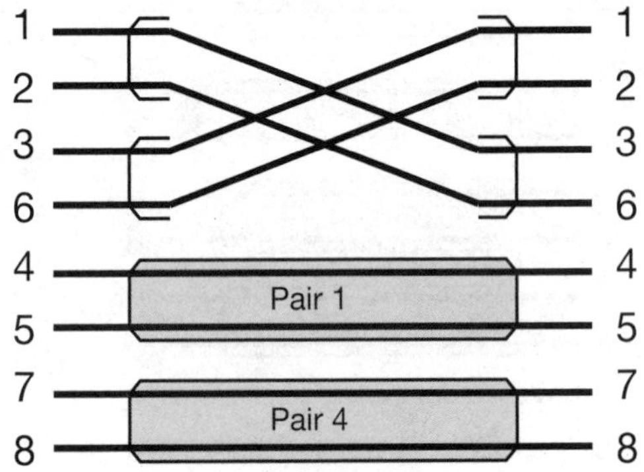

BALANCED PAIR TRANSMISSION

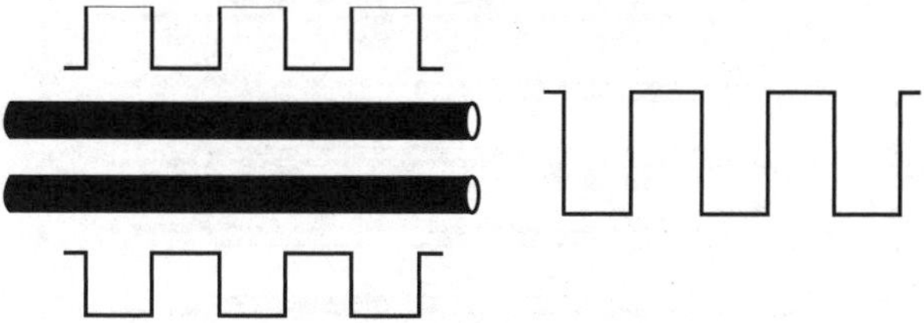

Equal but opposite signals on a pair of wires.

Output is sum of both signals.

ATTENUATION

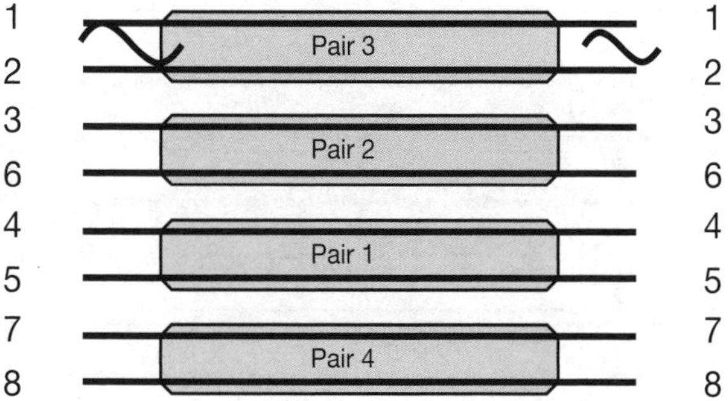

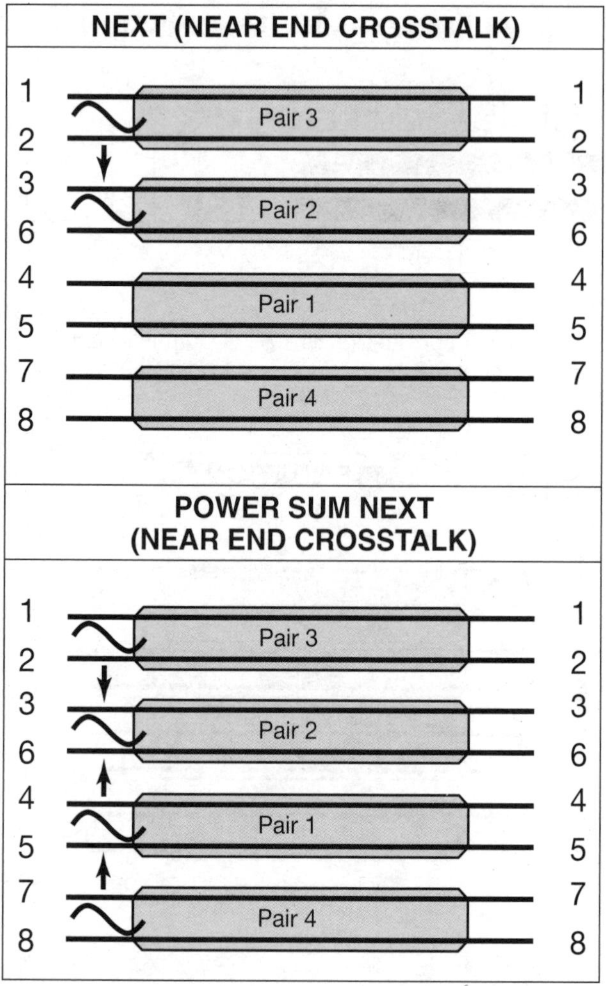
NEXT (NEAR END CROSSTALK)
1
2
3
6
4
5
7
8
Pair 3
Pair 2
Pair 1
Pair 4
1
2
3
6
4
5
7
8
POWER SUM NEXT
(NEAR END CROSSTALK)
1
2
3
6
4
5
7
8
Pair 3
Pair 2
Pair 1
Pair 4
1
2
3
6
4
5
7
8

ACR (ATTENUATION TO CROSSTALK RATIO)

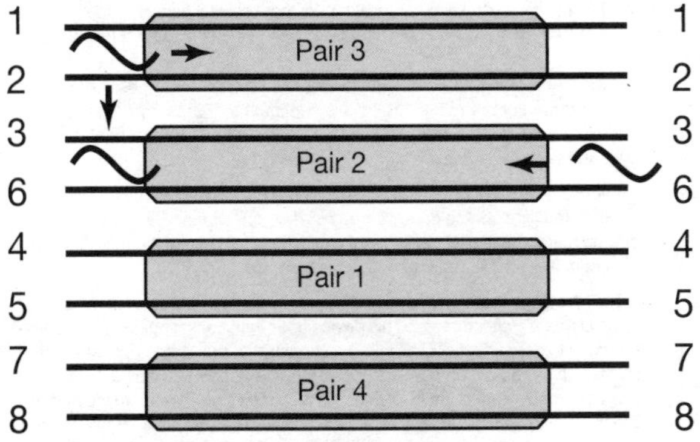

DELAY SKEW

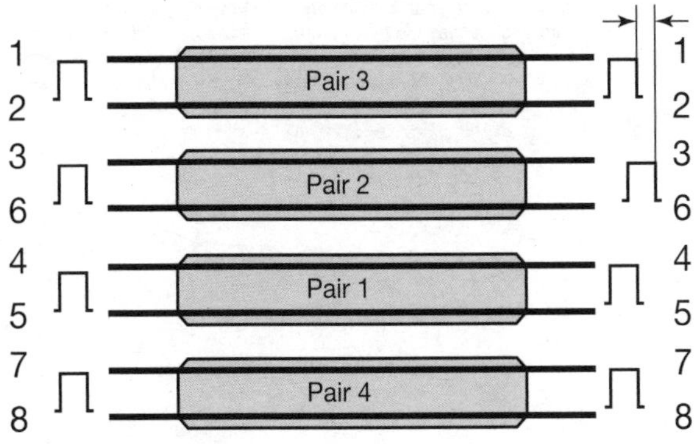

TESTING OPTICAL FIBER

When installing a fiber cable plant, it must be tested, to assure that it will properly perform. The purpose of testing is to make sure that light will pass through the system properly. The main types of optical testing used in the field are the following:

- **Continuity testing.** This is a simple visible light test. Its purpose is to make sure that the fibers in your cables are continuous, that is, that they are not broken. This is done with a modified type of flashlight device and the naked eye, and takes only a few minutes to perform.
- **Power testing.** This is to accurately measure the quality of optical fiber links. A calibrated light source puts infrared light into one end of the fiber, and a calibrated meter measures the light arriving at the other end of the fiber. The loss of light in the fiber is measured in decibels.
- **OTDR testing.** The OTDR is a piece of equipment properly called an *Optical Time Domain Reflectometer.* This device uses light backscattering to analyze fibers. In essence, the OTDR takes a snapshot of the fiber's optical characteristics. It sends a high powered pulse into the fiber and measures the light scattered back toward the instrument. The OTDR can be used to locate fiber breaks, splices and connectors, as well as to measure loss. However, the OTDR method of loss measurement may not give the same value for loss as a source and power meter, due to the different methods of measuring loss. In addition, the OTDR gives a graphic display of the status of the fiber being tested. Another advantage is that it requires access to only one end of the fiber.

  As useful as the OTDR is, however, it is not necessary in the majority of situations. A power meter and source are used to test the loss of fiber optic cable, simulating the way the fibers are used, and measuring the light lost from one end of the cable to the other. In addition, OTDRs are quite expensive. Even when they are necessary, many installers prefer to rent them, rather than purchasing them.

BASIC FEATURES OF THE OTDR TRACE

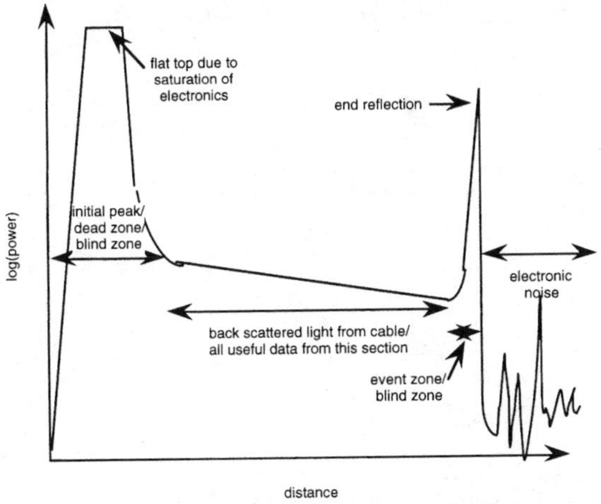

MICROSCOPE INSPECTION OF OPTICAL CONNECTORS

1. By viewing directly at the end of the polished surface with side lighting.
2. By viewing directly with side lighting and light transmitted through the core.
3. By viewing at an angle with lighting from the opposite angle.

Viewing directly with side lighting allows determining if the ferrule hole is of the proper size, the fiber is centered in the hole and a proper amount of adhesive has been applied. Only the largest scratches will be visible this way, however. Adding light transmitted through the core will make cracks in the end of the fiber, caused by pressure or heat during the polish process, visible.

Viewing the end of the connector at an angle, while lighting it from the opposite side at approximately the same angle will allow the best inspection for the quality of polish and possible scratches. The shadowing effect of angular viewing enhances the contrast of scratches against the mirror smooth polished surface of the glass.

One needs to be careful in inspecting connectors, however. The tendency is to be overly critical, especially at high magnification. Only defects over the fiber core are a problem. Chipping of the glass around the outside of the cladding is not unusual and will have no effect on the ability of the connector to couple light in the core. Likewise, scratches only on the cladding will not cause any loss problems.

INSPECTING FIBER CONNECTION WITH MICROSCOPE

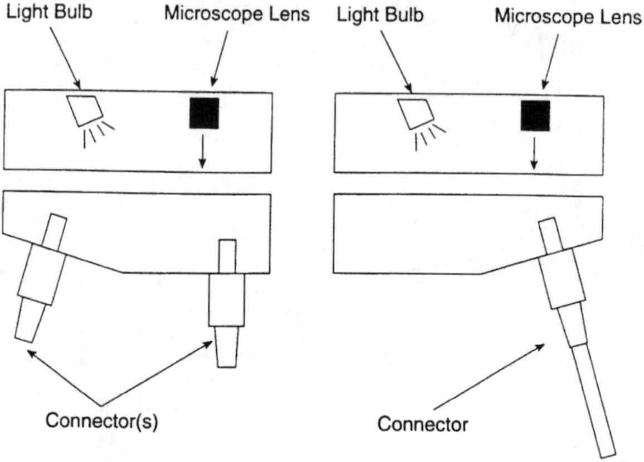

Direct View with Core Illumination

Angle View

FIBER OPTIC TESTING REQUIREMENTS

Test Parameter	Instrument
Optical power (source output, receiver signal lever)	Fiber optic power meter
Attenuation or loss of fibers, cables, and connectors	FO power meter and source, test kit or OLTS (optical loss test set)
Source wavelength*	FO spectrum analyzer
Backscatter (loss, length, fault location)	Optical time domain reflectometer (OTDR)
Fault location	OTDR, visual cable fault locator
Bandwidth/dispersion* (modal and chromatic)	Bandwidth tester or simulation software

*Rarely tested in the field

OPTICAL POWER LEVELS OF FIBER OPTIC COMMUNICATION SYSTEMS

Network Type	Wavelength (nm)	Power Range (dBm)	Power Range (W)
Telecom	1300, 1550	+3 to − 45	50 nW to 2mW
Datacom	665, 790, 850, 1300	−10 to −30	1 to 100uW
CATV	1300, 1550	+10 to −6	250uW to 10mW

CHARACTERISTICS OF DETECTORS USED IN FIBER OPTIC POWER METERS

Detector Type	Wavelength Range (nm)	Power Range (dBm)	Comments
Silicon	400–1100	+10 to −70	
Germanium	800–1600	+10 to −60	−70 with small area detectors, +30 with attenuator windows
InGaAs	800–1600	+10 to −70	Small area detectors may overload at high power (>.0 dBm)

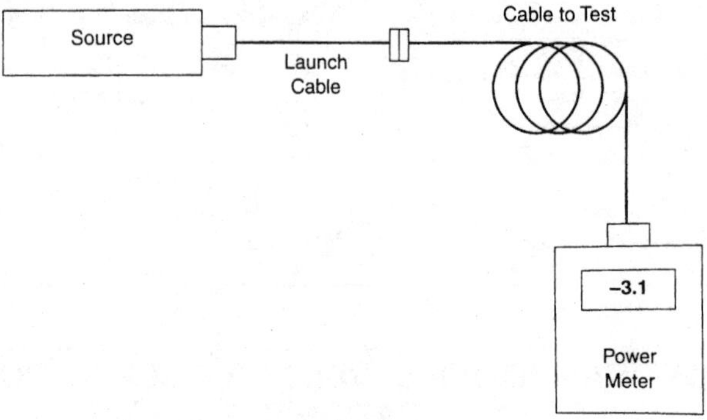
Source
Launch
Cable
Cable to Test
-3.1
Power
Meter

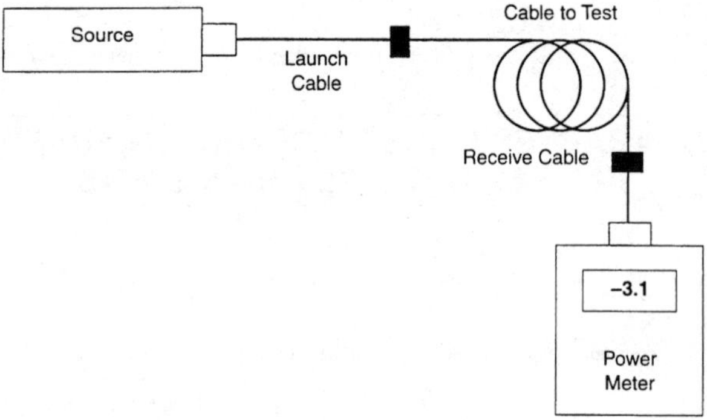
Source
Launch
Cable
Cable to Test
Receive Cable
-3.1
Power
Meter

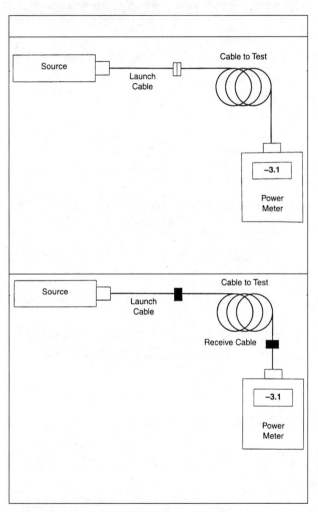

MISMATCHED FIBER CONNECTION LOSSES (EXCESS LOSS IN dB)

	Transmitting Fiber		
Receiving Fiber	62.5/125	85/125	100/140
50/125	0.9–1.6	3.0–4.6	4.7–9
62.5/125	—	0.9	2.1–4.1
85/125	—	—	0.9–1.4

OFSTP-14 CABLE PLANT LOSS TEST

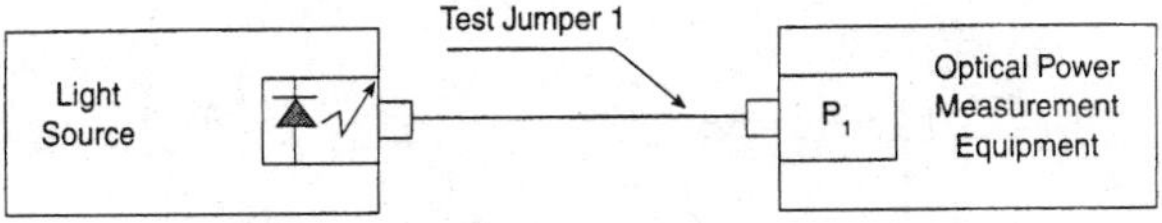

Reference Power Measurement for Method B

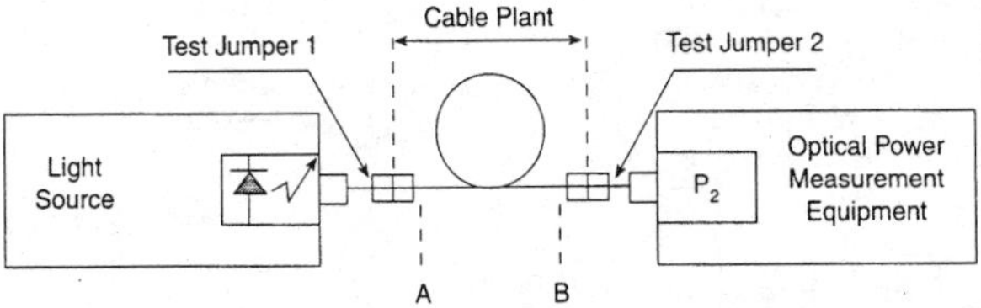

TYPICAL FIBER OPTIC LINK

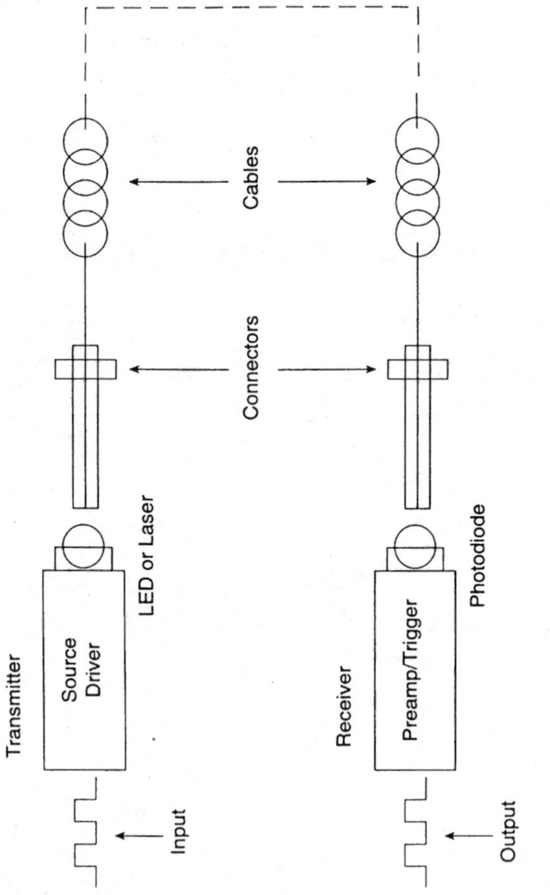

FIBER OPTIC LOSS GUIDELINES

- **Connector Loss.** For each connector, figure 0.5 dB loss (0.75 max per EIA/TIA 568)
  When testing cable plants per OFSTP-14 (double ended), include connectors on both ends of the cable. When testing per FOTP-171 (single ended), include only one connector - the one attached to the launch cable.

- **Splice Loss.** For each splice, figure 0.2 dB (0.3 max per EIA/TIA 568)

- **Fiber Loss** For multimode fiber, the loss is about 3 dB per km for 850 nm sources, 1 dB per km for 1300 nm. (3.5 and 1.5 dB/km max per EIA/TIA 568) This roughly translates into a loss of 0.1 dB per 100 feet (30 m) for 850 nm, 0.1 dB per 300 feet(100 m) for 1300 nm.
  For singlemode fiber, the loss is about 0.5 dB per km for 1300 nm sources, 0.4 dB per km for 1550 nm. (1.0 dB/km for premises/0.5 dB/km at either wavelength for outside plant max per EIA/TIA 568) This roughly translates into a loss of 0.1 dB per 600 (200m) feet for 1300 nm, 0.1 dB per 750 feet (250m) for 1300 nm.

So for the estimated loss of a cable plant:
Calculate the appropriate loss as:

(0.5 dB x # connectors) + (0.2 dB x # splices) + (fiber attenuation X the total length of cable)

EIA 445 FIBER OPTIC TEST PROCEDURES (FOTPs)

FOTP - 1	Cable Flexing for Fiber Optic Interconnecting Devices
FOTP - 2	Impact Test Measurements for Fiber Optic Devices
FOTP - 3	Temperature Effects Measurement Procedure for Optical Fiber, Optical Cable, and Other Passive Components
FOTP - 4	Fiber Optic Connector/Component Temperature Life
FOTP - 5	Humidity Test Procedure for Fiber Optic Connecting Devices
FOTP - 6	Cable Retention Test Procedure for Fiber Optic Cable Interconnecting Devices
FOTP - 10	Measurement of the Amount of Extractable Material in Coatings Applied to Optical Fiber
FOTP - 11	Vibration Test Procedure for Fiber Optic Connecting Devices and Cables
FOTP - 12	Fluid Immersion Test for Fiber Optic Components
FOTP - 13	Visual and Mechanical Inspection of fibers, Cables, Connectors, and Other Devices
FOTP - 14	Physical Shock (Specified Pulse)
FOTP - 15	Altitude Immersion

EIA 445 FIBER OPTIC TEST PROCEDURES (FOTPs) *(cont'd)*

FOTP - 16	Salt Spray (Corrosion) Test for Fiber Optic Components
FOTP - 17	Maintenance Aging of Fiber Optic Connectors and Terminated Cable Assemblies
FOTP - 18	Acceleration Testing for Components and Assemblies
FOTP - 20	Measurement of Change in Optical Transmittance
FOTP - 21	Mating Durability for Fiber Optic Interconnecting Devices
FOTP - 22	Ambient Light Susceptibility of Components
FOTP - 23	Air Leakage Testing for fiber Optic Component Seals
FOTP - 24	Water Peak Attenuation Measurement of Single-mode Fibers
FOTP - 25	Repeated Impact Testing of Fiber Optic Cables and Cable Assemblies
FOTP - 26	Crush Resistance of Fiber Optic Interconnecting Devices
FOTP - 27	Fiber Diameter Measurements
FOTP - 28	Measurement of Dynamic Tensile Strength of Optical Fiber
FOTP - 29	Refractive index Profile (Transverse Interference Method)
FOTP - 30	Frequency Domain Measurement of Multimode Optical Fiber Information Transmission Capacity
FOTP - 31	Fiber Tensile Proof Test Method
FOTP - 32	Fiber Optic Circuit Discontinuities
FOTP - 33	Fiber Optic Cable Tensile Loading and Bending Test
FOTP - 34	Interconnection Device Insertion Loss Test
FOTP - 35	Fiber Optic Component Dust (Fine Sand) Test
FOTP - 36	Twist Test for Connecting Devices
FOTP - 37	Fiber Optic Cable Bend Test, Low and High Temperature
FOTP - 39	Water Wicking Test for Fiber Optic Cable
FOTP - 40	Fluid Immersion, Cables
FOTP - 41	Compressive Loading Resistance of Fiber Optic Cables
FOTP - 42	Optical Crosstalk in Components
FOTP - 43	Output Near field Radiation Pattern Measurement of Optical Waveguide Fibers
FOTP - 44	Refractive Index Profile (Refracted Ray Method)
FOTP - 45	Microscopic Method for Measuring Fiber Geometry of Optical Waveguide Fibers
FOTP - 46	Spectral Attenuation Measurement (Long Length Graded Index Optical Fibers)
FOTP - 47	Output Far Field Radiation Pattern Measurement
FOTP - 48	Measurement of Optical Fiber Cladding Diameter Using Laser-Based Instruments
FOTP - 49	Measurement for Gamma Irradiation Effects on Optical Fiber and Cables
FOTP - 50	Light Launch Conditions (Long Length Graded Index Fibers)
FOTP - 51	Pulse Distortion Measurement, Multimode Fiber
FOTP - 53	Attenuation by Substitution (Multimode Graded Index)

EIA 445 FIBER OPTIC TEST PROCEDURES (FOTPs) *(cont'd)*

FOTP - 54	Mode Scrambler Requirements for Overfilled Launching Conditions (Multimode)
FOTP - 55	End View Methods for Measuring Coating and Buffer Geometry
FOTP - 56	Test Method for Evaluating Fungus Resistance of Optical Waveguide Fibers and Cables
FOTP - 57	Optical Fiber End Preparation and Examination
FOTP - 58	Core Diameter Measurements (Graded Index Fibers)
FOTP - 59	Measurement of Fiber Point Defects Using an OTDR
FOTP - 60	Measurement of Fiber or Cable Length Using an OTDR
FOTP - 61	Measurement of Fiber or Cable Attenuation Using an OTDR
FOTP - 62	Optical Fiber Macrobend Attenuation
FOTP - 63	Torsion Test for Optical Fiber
FOTP - 65	Flexure Test for Optical Fiber
FOTP - 66	Test Method for Measuring Relative Abrasion Resistance
FOTP - 68	Optical Fiber Microbend Test Procedure
FOTP - 69	Evaluation of Minimum and Maximum Exposure Temperature on the Optical Performance of Optical Fiber
FOTP - 71	Measurement of Temperature Shock Effects on Components
FOTP - 75	Fluid Immersion Test for Optical Waveguide Fibers
FOTP - 77	Procedure to Qualify a Higher-Order Mode Filter for Measurements of Single-mode Fibers
FOTP - 78	Spectral Attenuation Cutback Measurement (Single-mode)
FOTP - 80	Cutoff Wavelength of Uncabled Single-mode Fiber by Transmitted Power
FOTP - 81	Compound Flow (Drip) Test for Filled Fiber Optic Cable
FOTP - 82	Fluid Penetration Test for Fluid-Blocked Cable
FOTP - 83	Cable to Interconnecting Device Axial Compressive Loading
FOTP - 84	Jacket Self-Adhesion (Blocking) Test for Cables
FOTP - 85	Fiber Optic Cable Twist Test
FOTP - 86	Fiber Optic Cable Jacket Shrinkage
FOTP - 87	Fiber Optic Cable Knot Test
FOTP - 88	Fiber Optic Cable Bend Test
FOTP - 89	Fiber Optic Cable Jacket Elongation and Tensile Strength Test
FOTP - 91	Fiber Optic Cable Twist-Bend Test
FOTP - 92	Optical Fiber Cladding Diameter and Noncircularity Measurement by Fizeau Interferometry
FOTP - 94	Fiber Optic Cable Stuffing Tubing Compression
FOTP - 95	Absolute Optical Power Test for Fibers and Cables
FOTP - 96	Fiber Optic Cable Long-Term Storage Temperature Test for Extreme Environments
FOTP - 98	Fiber Optic Cable External Freezing Test
FOTP - 99	Gas Flame Test for Special Purpose Cable
FOTP - 100	Gas Leakage Test for Gas Blocked Cable
FOTP - 101	Accelerated Oxygen Test

EIA 445 FIBER OPTIC TEST PROCEDURES (FOTPs) *(cont'd)*

FOTP - 102	Water Pressure Cycling
FOTP - 104	Fiber Optic Cable Cyclic Flexing Test
FOTP - 107	Return Loss for Fiber Optic Components
FOTP - 127	Spectral Characterization of Multimode Laser Diodes
FOTP - 162	Fiber Optic Cable Temperature-Humidity Cycling
FOTP - 164	Measurement of Mode Field Diameter by Far-Field Scanning (Single-mode)
FOTP - 165	Measurement of Mode Field Diameter by Near Field Scanning (Single-mode)
FOTP - 166	Transverse Offset Method
FOTP - 167	Mode Field Diameter Measurement, Variable Aperture Method in Far-Field
FOTP - 168	Chromatic Dispersion Measurement of Multimode Graded-Index and Single-mode Optical fiber by Phase-Shift Method
FOTP - 169	Chromatic Dispersion Measurement of Optical Fibers by the Phase-Shift Method
FOTP - 170	Cable Cutoff Wavelength of Single-mode Fiber by Transmitted Power
FOTP - 171	Attenuation by Substitution Measurement (Short Length Multi-mode Graded-Index and Single-mode
FOTP - 172	Flame Resistance of Firewall Connector
FOTP - 173	Coating Geometry Measurement of Optical Fiber, Side-View Method
FOTP - 174	Mode Field Diameter of Single-mode Fiber by Knife-Edge Scanning in Far-Field
FOTP - 175	Chromatic Dispersion Measurement of Optical Fiber by the Differential Phase-Shift
FOTP - 177	Numerical Aperture Measurement of Graded-Index Fiber
FOTP - 178	Coating Strip Force Measurement
FOTP - 179	Inspection of Cleaved Fiber End Faces by Interferometry
FOTP - 180	Measurement of Optical Transfer Coefficients of a Passive Branching Device
FOTP - 184	Coupling Proof Overload Test for Fiber Optic Interconnecting Devices
FOTP - 185	Strength of Coupling Mechanism for Fiber Optic Interconnecting Devices
FOTP - 186	Gauge Retention Force Measurement for Components
FOTP - 187	Engagement and Separation/Force for Connector Sets
FOTP - 188	Low-Temperature Testing for Components
FOTP - 189	Ozone Exposure Test for Components
FOTP - 190	Low Air Pressure (High Altitude) Test for Components
FOTP - 191	Measurement of Mode Field Diameter of Single-Mode Optical fiber

EIA 445 FIBER OPTIC TEST PROCEDURES (FOTPs) *(cont'd)*

FOTP - 193	Polarization Crosstalk Method for Polarization Maintaining Optical Fiber and Components
FOTP - 195	Coating Geometry Measurement for Optical Fiber
FOTP - 196	Guideline For Polarization-Mode Measurement In Single Mode Fiber Optic Components and Devices
FOTP - 197	Differential Group Delay Measurement Of Single-Mode Components and Devices by the Differential Phase Shift Method
FOTP - 200	Insertion Loss of Connectorized Polarization-Maintaining Fiber or Polarizing Fiber Pigtailed Devices and Cable Assemblies
FOTP - 201	Return Loss of Connectorized Polarization-Maintaining Fiber or Polarizing Fiber Pigtailed Devices and Cable Assemblies
FOTP - 203	(Transceiver Output Pattern)
FOTP - 204	(Restricted Modal Launch Bandwidth for Multimode Fibers)
FOTP - 220	(Differential Modal Dispersion in Multimode Fiber - Pulse Broadening as a Function of Source Launch Offsets)

SYSTEM TEST PROCEDURES

OFSTP - 10	Measurement of Dispersion Power Penalty In Digital Single-Mode Systems
OFSTP - 11	Measurement of Single-Reflection Power Penalty for Fiber Optic Terminal Equipment
OFSTP - 14	Optical Power Loss Measurements of installed Multimode Fiber Cable Plant
OFSTP - 15	Jitter Tolerance Measurement
OFSTP - 16	Jitter Transfer Function Measurement
OFSTP - 17	Output Jitter Measurement
OFSTP - 18	Systematic Jitter Generation Measurement
OFSTP - 19	Optical Signal-To-Noise Ratio Measurement Procedures for Dense Wavelength-Division Multiplexed Systems
OFSTP - 27	Procedure for System Level Temperature Cycle Endurance Test
OFSTP - 28	IEC - 61290 - 1- 2: Basic Spec for Optical Fiber Amplifiers Test Methods Part 1: Test Methods for Gain Parameters - Sect. 2: Electrical Spectrum Analyzer Test Method
OFSTP - 29	IEC - 61290 - 1- 3: Basic Spec for Optical Fiber Amplifiers Test Methods Part 1: Test Methods for Gain Parameters - Sect. 3: Optical Power Meter Test Method
OFSTP - 30	IEC - 61290 - 2-1: Basic Specification For Optical Fiber Amplifiers Test Methods - part 2: Test Methods For Spectral Power Parameters - Section 2 - Optical Spectrum Analyzer Test Method

EIA 458-B STANDARD OPTICAL FIBER MATERIAL CLASSES AND PREFERRED SIZES

EIA - 472	General Specification for Fiber Optic Cable
EIA - 472A	Sectional Specification for Fiber Optic Communication Cables for Outside Aerial Use
EIA - 472B	Sectional Specification for Fiber Optic Communication Cables for Underground and Buried Use
EIA - 472C	Sectional Specification for Fiber Optic Communication Cables for Indoor Use
EIA - 472D	Sectional Specification for Fiber Optic Communication Cables for Outside Telephone Plant Use
EIA - 4750000 - B	Generic Specification for Fiber Optic Connectors
EIA - 475C000	Sectional Specification for Type FSMA Connectors
EIA - 475CA000	Blank Detail Specification for Optical Fiber and Cable Type FSMA, Environmental Category I EIA - 475CBOO Blank Detail Specification Connector Set for Optical Fiber and Cables Type FSMA, Environmental Category II
EIA - 475CC000	Blank Detail Specification Connector Set for Optical Fiber and Cables Type FSMA, Environmental Category III
EIA - 475E0000	Sectional Specification for Fiber Optic Connectors Type BFOC/2.5
EIA - 475EA00	Blank Detail Specification for Connector Set for Optical Fiber and Cables, Type BFOC/2.5, Environmental Category I
EIA - 475EB00	Blank Detail Specification for Connector Set for Optical Fiber and Cables, Type BFOC/2.5, Environmental Category II
EIA - 475EC00	Blank Detail Specification for Connector Set for Optical Fiber and Cables, Type BFOC/2.5, Environmental Category III
EIA - 492AAAA	Detail Specification for 62.5 micron Core Diameter/l 25 micron Cladding Diameter Class I A Multi-node, Graded Index Optical Waveguide Fibers
EIA - 5390000	Generic Specification for Field Portable Polishing Device for Preparation Optical Fiber
EIA - 5460000	Generic Specification for a Field Portable Optical Inspection Device, Combined EIA - NECQ Specification
EIA - 546A000	Sectional Specification for a Field Portable Optical Microscope for Inspection of Optical Waveguide and Related Devices
EIA - 587	Fiber Optic Graphic Symbols
EIA - 590	Standard for Physical Location and Protection of Below-Ground Fiber Optic Cable Plant
EIA - 598	Color Coding of Fiber Optic Cables

CHAPTER 6
TELEDATA

TELEPHONE CIRCUIT OPERATION

VOICE NETWORKS

- **Traditional Telephone Network**
  - Subscriber (business, residential)
  - LEC (central office)
  - Long Distance Carrier
- **Business Phone Systems**
  - Multiline Phone (KSUless)
  - Key System
  - PBX
- **Centrex**
  - Provided by the LEC
  - Can coincide with a PBX or KSU
  - Utilizes standard telephone sets
- **ISDN**
  - LEC must have compatible equipment
  - Digital technology
  - Interaction with PC network

WIRING CLOSETS

- **Definition:**
  - an area that provides safe and environmentally suitable installation of cables, wires, cross-connect fields and premise equipment. Usually associated with one floor of a building.
- **Installation Standards/Regulations**
  - ANSI/EIA/TIA 568—ANSI/NFPA 70
  - · · · 569—FCC Part 68
  - · · · 570—UL 1459
  - · · · 606—UL 1863
  - · · · 607
- **Design Considerations**
  - location
  - security
  - bonding and grounding
  - fire protection

EQUIPMENT ROOMS

- **Definition**
  - a specialized room that provides space for and maintains a suitable operating environment for voice and data equipment. Usually larger than a "wiring closet" and may contain a work area for technicians.
- **Standards**
  - EIA/TIA 568
  - EIA/TIA 569
- **Design Considerations**
  - types of equipment to be placed
  - space
  - electrical specifications
  - structural
  - environmental

RATING PAIRED COPPER BUILDING CABLES

- **Fire Ratings . . . UL**
  1 General Purpose . . . . Good
  2 Riser Grade . . . . Better
  3 Plenum Grade . . . . Best

- **Performance Ratings**
  - Cat 1 . . . . Not Rated . . . . Usually Residential
  - Cat 2 . . . . up to 1 MHz . . . . Not for Horizontal
  - Cat 3 . . . . up to 10 Mb/s . . . . Typical Voice/Data
  - Cat 4 . . . . up to 16 Mb/s . . . . Seldom Used
  - Cat 5 . . . . up to 100 Mb/s . . . . Emerging Applications

HORIZONTAL FOUR-PAIR CAT 5 CABLING

- **Characteristics**
  - Twisted pair copper
  - 4 pairs
  - Rated to 100 MHz
  - Maximum length 259 feet from telecom closet to outlet.
  - Patch cords & cross-connects may add 20 feet.

TRADITIONAL TELEPHONE NETWORK (POTS)

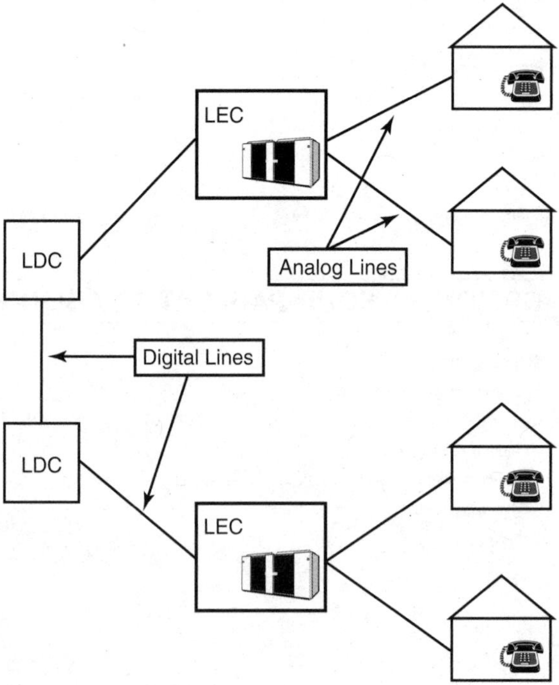

KEY SYSTEM

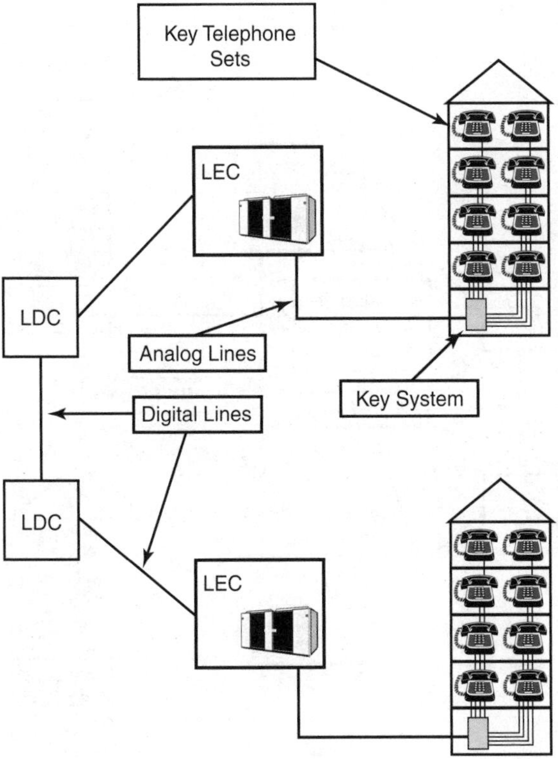

Key Telephone Sets
LEC
LDC
Analog Lines
Key System
Digital Lines
LDC
LEC

PRIVATE BRANCH EXCHANGE (PBX)

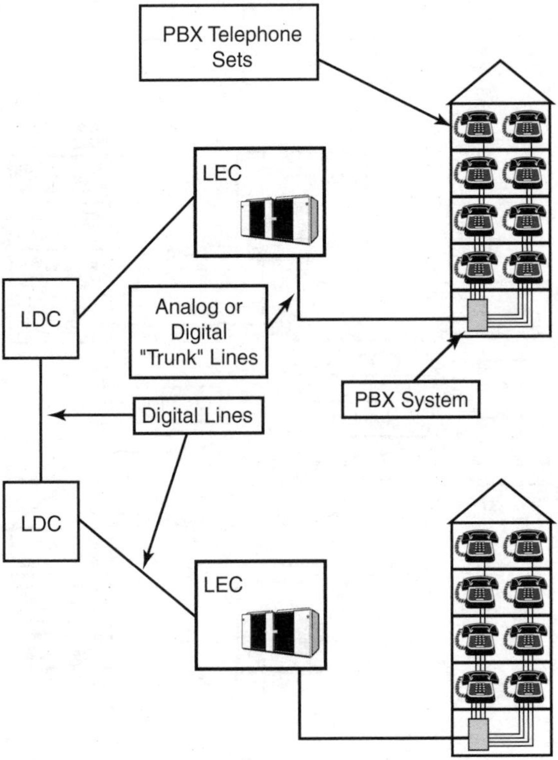

CENTREX

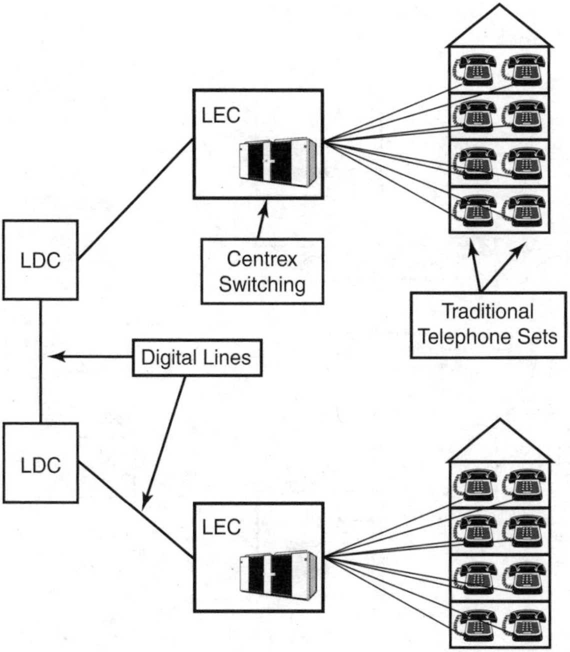

LEC
Centrex Switching
Traditional Telephone Sets
LDC
Digital Lines
LDC
LEC

INTEGRATED SERVICES DIGITAL NETWORK (ISDN)

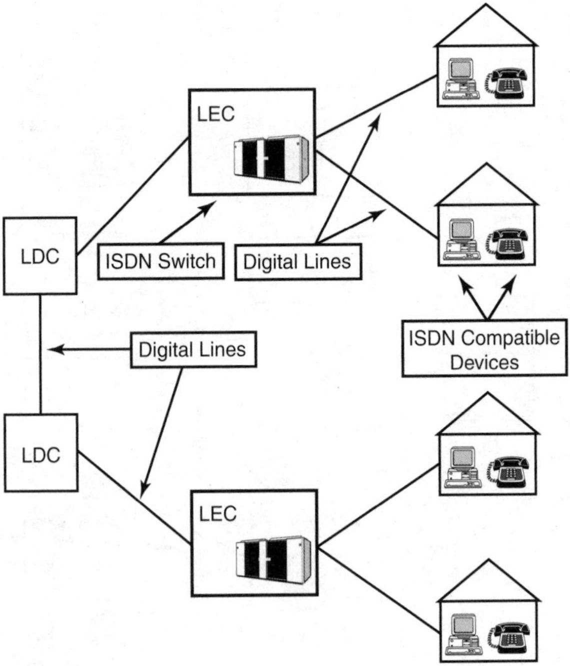

COMMON TELEPHONE CONNECTIONS

The most common and simplest type of communication installation is the single line telephone. The typical telephone cable (sometimes called quad cable) contains four wires, colored green, red, black, and yellow. A one line telephone requires only two wires to operate. In almost all circumstances, green and red are the two conductors used. In a common four-wire modular connector, the green and red conductors are found in the inside positions, with the black and yellow wires in the outer positions.

As long as the two center conductors of the jack (again, always green and red) are connected to live phone lines, the telephone should operate.

Two-line phones generally use the same four wire cables and jacks. In this case, however, the inside two wires (green and red) carry line 1, and the outside two wires (black and yellow) carry line 2.

COLOR CODING OF CABLES

The color coding of twisted-pair cable uses a color pattern that identifies not only what conductors make up a pair but also what pair in the sequence it is, relative to other pairs within a multipair sheath. This is also used to determine which conductor in a pair is the *tip* conductor and which is the *ring* conductor. (The tip conductor is the positive conductor, and the ring conductor is the negative conductor.)

The banding scheme uses two opposing colors to represent a single pair. One color is considered the primary while the other color is considered the secondary. For example, given the primary color of white and the secondary color of blue, a single twisted-pair would consist of one cable that is white with blue bands on it. The five primary colors are white, red, black, yellow, and violet.

In multi-pair cables the primary color is responsible for an entire group of pairs (five pairs total). For example, the first five pairs all have the primary color of white. Each of the secondary colors, blue, orange, green, brown, and slate, are paired in a banded fashion with white. This continues through the entire primary color scheme for all four primary colors (comprising 25 individual pairs). In larger cables (50 pairs and up), each 25-pair group is wrapped in a pair of ribbons, again representing the groups of primary colors matched with their respective secondary colors. These color coded band markings help cable technicians to quickly identify and properly terminate cable pairs.

EIA COLOR CODE

You should note that the new EIA color code calls for the following color coding:

Pair 1 – White/Blue (white with blue stripe) and Blue
Pair 2 – White/Orange and Orange
Pair 3 – White/Green and Green
Pair 4 – White/Brown and Brown

TWISTED-PAIR PLUGS AND JACKS

One of the more important factors regarding twisted-pair implementations is the cable jack or cross-connect block. These items are vital since without the proper interface, any twisted-pair cable would be relatively useless. In the twisted-pair arena, there are three major types of twisted-pair jacks:

RJ-type connectors (phone plugs)

Pin-connector

Genderless connectors (IBM sexless data connectors)

The RJ-type (registered jack) name generally refers to the standard format used for most telephone jacks. The term pin-connector refers to twisted pair connectors, such as the RS-232 connector, which provide connection through male and female pin receptacles. Genderless connectors are connectors in which there is no separate male or female component; each component can plug into any other similar component.

STANDARD PHONE JACKS

The standard phone jack is specified by a variety of different names, such as RJ and RG, which refer to their physical and electrical characteristics. These jacks consist of a male and a female component. The male component snaps into the female receptacle. The important point to note, however, is the number of conductors each type of jack can support.

Common configurations for phone jacks include support for four, six, or eight conductors. A typical example of a four-conductor jack, supporting two twisted-pairs, would be the one used for connecting most telephone handsets to their receivers.

A common six-conductor jack, supporting three twisted-pairs, is the RJ-11 jack used to connect most telephones to the telephone company or PBX systems. An example of an eight-conductor jack Is the R-45 jack, which is intended for use under the ISDN system as the user-site interface for ISDN terminals.

For building wiring, the six-conductor and eight-conductor jacks are popular, with the eight-conductor jack increasing in popularity, as more corporations install twisted-pair in four-pair bundles for both voice and data. The eight-conductor jack, in addition to being used for ISDN, is also specified by several other popular applications, such as the new IEEE 802.3 10 BaseT standard for Ethernet over twisted-pair.

These types of jacks are often keyed, so that the wrong type of plug cannot be inserted into the jack. There are two kinds of keying -- side keying, and shift keying.

Side keying uses a piece of plastic that is extended to one side of the jack. This type is often used when multiple jacks are present.

Shift keying entails shifting the position of the snap connector to the left or right of the jack, rather than leaving it in its usual center position. Shift keying Is more commonly used for data connectors than for voice connectors.

Note that while we say that these jacks are used for certain types of systems (data, voice, etc.) this is not any type of standard. They can be used as you please.

PIN CONNECTORS

There are any number of pin type connectors available. The most familiar type Is the RS-232 jack that is commonly used for computer ports. Another popular type of pin connector is the DB type connector, which is the round connector that is commonly used for computer keyboards.

The various types of pin connectors can be used for terminating as few as five (the DB type), or more than 50 (the RS type) conductors.

50-pin *champ* type connectors are often used with twisted-pair cables, when connecting to cross-connect equipment, patch panels, and communications equipment such as Is used for networking.

CROSS CONNECTIONS

Cross connections are made at terminal *blocks*. A block is typically a rectangular, white plastic unit, with metal connection points. The most common type is called a punch-down block. This is the kind that you see on the back wall of a business, where the main telephone connections are made.The wire connections are made by pushing the insulated wires into their places. When "punched" down, the connector cuts through the insulation, and makes the appropriate connection.

Connections are made between punch-down blocks by using *patch cords*, which are short lengths of cable that can be terminated into the punch-down slots, or that are equipped with connectors on each end.

When different systems must be connected together, cross-connects are used.

CATEGORY CABLING

Category 1 cable is the old standard type of telephone cable, with four conductors colored green, red, black, and yellow. Also called quad cable.

Category 2 was an old IBM cabling system. It is almost never used for modern communications.

Category 3 cable is used for digital voice and data transmlsslon rates up to 10 Mbps (Megabits per second). Common types of data transmission over this communications cable would be UTP Token Ring (4 Mbps) and 10Base-T (10 Mbps).

Category 4 cable is used for voice and data transmission rates up to 16 Mbps. A typical type of transmission over this system would be UTP Token Ring (16 Mbps).

Category 5 cable is used for sending voice and data at speeds up to 100Mbit/s (megabits per second).

This includes signals used under the FDDI communications standard.

INSTALLATION REQUIREMENTS

Article 800 of the NEC covers communication circuits, such as telephone systems and outside wiring for fire and burglar alarm systems. Generally these circuits must be separated from power circuits and grounded. In addition, all such circuits that run out of doors (even if only partially) must be provided with circuit protectors (surge or voltage suppressors).

The requirements for these installations are as follows:

CONDUCTORS ENTERING BUILDINGS

If communications and power conductors are supported by the same pole, or run parallel in span, the following conditions must be met :

1. Wherever possible, communications conductors should be located below power conductors.
2. Communications conductors cannot be connected to crossarms .
3. Power service drops must be separated from communications service drops by at least 12 inches.

Above roofs, communications conductors must have the following clearances:

1. Flat roofs: 8 feet.
2. Garages and other auxiliary buildings: None required.
3. Overhangs, where no more than 4 feet of communications cable will run over the area: 18 inches.
4. Where the roof slope is 4 inches rise for every 12 inches horizontally: 3 feet.

Underground communications conductors must be separated from power conductors in manhole or handholes by brick, concrete, or tile partitions.

Communications conductors should be kept at least 6 feet away from lightning protection system conductors.

CIRCUIT PROTECTION

Protectors are surge arresters designed for the specific requirements of communications circuits. They are required for all aerial circuits not confined with a *block*. (Block here means city block.) They must be installed on all circuits with a block that could accidentally contact power circuits over 300 volts to ground. They must also be listed for the type of installation.

Other requirements are the following:

Metal sheaths of any communications cables must be grounded or interrupted with an insulating joint as close as practicable to the point where they enter any building (such point of entrance being the place where the communications cable emerges through an exterior wall or concrete floor slab, or from a grounded rigid or intermediate metal conduit).

Grounding conductors for communications circuits must be copper or some other corrosion-resistant material, and have insulation suitable for the area in which it is installed.

Communications grounding conductors may be no smaller than No. 14.

The grounding conductor must be run as directly as possible to the grounding electrode, and be protected if necessary.

If the grounding conductor Is protected by metal raceway, it must be bonded to the grounding conductor on both ends.

CIRCUIT PROTECTION *(cont'd)*

Grounding electrodes for communications ground may be any of the following:
1. The grounding electrode of an electrical power system.
2. A grounded interior metal piping system (Avoid gas piping systems for obvious reasons.)
3. Metal power service raceway.
4. Power service equipment enclosures.
5. A separate grounding electrode.

If the building being served has no grounding electrode system, the following can be used as a grounding electrode:
1. Any acceptable power system grounding electrode. (See Section 250-81.)
2. A grounded metal structure.
3. A ground rod or pipe at least 5 feet long and 1/2 inch in diameter. This rod should be driven into damp (if possible) earth, and kept separate from any lightning protection system grounds or conductors.

Connections to grounding electrodes must be made with approved means .

If the power and communications systems use separate grounding electrodes, they must be bonded together with a No. 6 copper conductor. Other electrodes may be bonded also. This is not required for mobile homes.

For mobile homes, if there is no service equipment or disconnect within 30 feet of the mobile home wall, the communications circuit must have its own grounding electrode. In this case, or if the mobile home is connected with cord and plug, the communications circuit protector must be bonded to the mobile home frame or grounding terminal with a copper conductor no smaller than No. 12.

INTERIOR COMMUNICATIONS CONDUCTORS

Communications conductors must be kept at least 2 inches away from power or Class 1 conductors, unless they are permanently separated from them or unless the power or Class 1 conductors are enclosed in one of the following:
1. Raceway.
2. Type AC, MC, UF, NM, or NM cable, or metal-sheathed cable.

Communications cables are allowed in the same raceway, box, or cable with any of the following:
1. Class 2 and 3 remote-control, signaling, and power-limited circuits.
2. Power-limited fire protective signaling systems.
3. Conductive or nonconductive optical fiber cables.
4. Community antenna television and radio distribution systems.

Communications conductors are not allowed to be in the same raceway or fitting with power or Class 1 circuits.

Communications conductors are not allowed to be supported by raceways unless the raceway runs directly to the piece of equipment the communications circuit serves.

Openings through fire-resistant floors, walls, etc. must be sealed with an appropriate firestopping material.

Any communications cables used in plenums or environmental air-handling spaces must be listed for such use.

STANDARD TELECOM COLOR CODING

PAIR #	TIP (+) COLOR	RING (–) COLOR
1	White	Blue
2	White	Orange
3	White	Green
4	White	Brown
5	White	Slate
6	Red	Blue
7	Red	Orange
8	Red	Green
9	Red	Brown
10	Red	Slate
11	Black	Blue
12	Black	Orange
13	Black	Green
14	Black	Brown
15	Black	Slate
16	Yellow	Blue
17	Yellow	Orange
18	Yellow	Green
19	Yellow	Brown
20	Yellow	Slate
21	Violet	Blue
22	Violet	Orange
23	Violet	Green
24	Violet	Brown
25	Violet	Slate

25-PAIR COLOR CODING/ISDN CONTACT ASSIGNMENTS

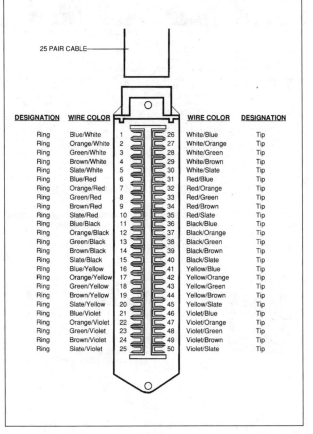

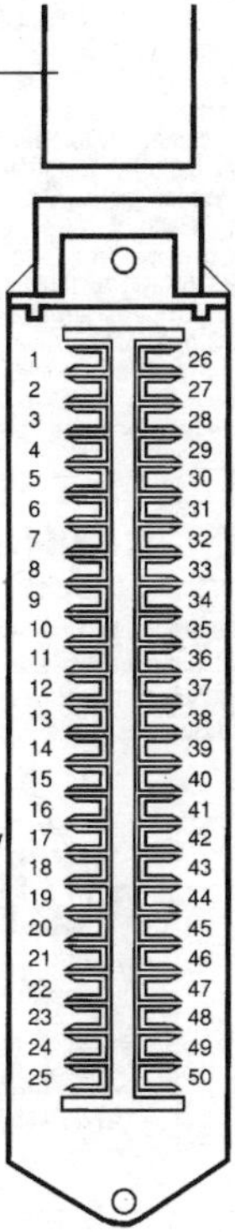

DESIGNATION	WIRE COLOR			WIRE COLOR	DESIGNATION
Ring	Blue/White	1	26	White/Blue	Tip
Ring	Orange/White	2	27	White/Orange	Tip
Ring	Green/White	3	28	White/Green	Tip
Ring	Brown/White	4	29	White/Brown	Tip
Ring	Slate/White	5	30	White/Slate	Tip
Ring	Blue/Red	6	31	Red/Blue	Tip
Ring	Orange/Red	7	32	Red/Orange	Tip
Ring	Green/Red	8	33	Red/Green	Tip
Ring	Brown/Red	9	34	Red/Brown	Tip
Ring	Slate/Red	10	35	Red/Slate	Tip
Ring	Blue/Black	11	36	Black/Blue	Tip
Ring	Orange/Black	12	37	Black/Orange	Tip
Ring	Green/Black	13	38	Black/Green	Tip
Ring	Brown/Black	14	39	Black/Brown	Tip
Ring	Slate/Black	15	40	Black/Slate	Tip
Ring	Blue/Yellow	16	41	Yellow/Blue	Tip
Ring	Orange/Yellow	17	42	Yellow/Orange	Tip
Ring	Green/Yellow	18	43	Yellow/Green	Tip
Ring	Brown/Yellow	19	44	Yellow/Brown	Tip
Ring	Slate/Yellow	20	45	Yellow/Slate	Tip
Ring	Blue/Violet	21	46	Violet/Blue	Tip
Ring	Orange/Violet	22	47	Violet/Orange	Tip
Ring	Green/Violet	23	48	Violet/Green	Tip
Ring	Brown/Violet	24	49	Violet/Brown	Tip
Ring	Slate/Violet	25	50	Violet/Slate	Tip

MODULAR JACK STYLES

8-Position

8-Position Keyed

6-Position

6-Position Modified

There are four basic modular jack styles. The 8-position and 8-position keyed modular jacks are commonly and incorrectly referred to as RJ45 and keyed RJ45 (respectively). The 6-position modular jack is commonly referred to as RJ11. Using these terms can sometimes lead to confusion since the RJ designations actually refer to very specific wiring configurations called Universal Service Ordering Codes (USOC). The designation 'RJ' means Registered Jack. Each of these 3 basic jack styles can be wired for different RJ configurations. For example, the 6-position jack can be wired as a RJ11C (1-Pair), RJ14C (2-Pair), or RJ25C (3-Pair) configuration. An 8-position jack can be wired for configurations such as RJ61C (4-Pair) and RJ48C. The keyed 8-position jack can be wired for RJ45S, RJ46S and RJ47S. The fourth modular jack style is a modified version of the 6-position jack (modified modular jack or MMJ). It was designed by DEC along with the modified modular plug (MMP) to eliminate the possibility of connecting DEC data equipment to voice lines and vice versa.

COMMON WIRING CONFIGURATIONS

The TIA and AT&T wiring schemes are the two that have been adopted by EIA/TIA-568. They are nearly identical except that pairs two and three are reversed. TIA is the preferred scheme because it is compatible with 1 or 2-pair USOC Systems. Either configuration can be used for Integrated Services Digital Network (ISDN) applications.

TIA (T568A)

Pair ID	PIN #
T1	5
R1	4
T2	3
R2	6
T3	1
R3	2
T4	7
R4	8

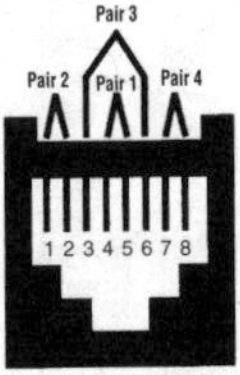

AT&T (T568B)

Pair ID	PIN #
T1	5
R1	4
T2	1
R2	2
T3	3
R3	6
T4	7
R4	8

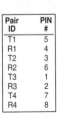

COMMON WIRING CONFIGURATIONS (*cont'd*)

USOC wiring is available for 1-, 2-, 3-, or 4-pair systems. Pair 1 occupies the center conductors, pair 2 occupies the next two contacts out, etc. One advantage to this scheme is that a 6-position plug configured with 1, 2, or 3 pairs can be inserted into an 8-position jack and maintain pair continuity; a note of warning though, pins 1 and 8 on the jack may become damaged from this practice. A disadvantage is the poor transmission performance associated with this type of pair sequence.

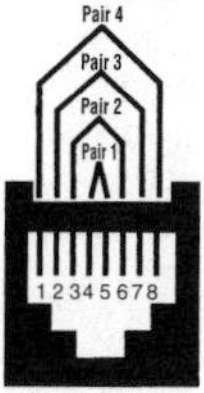

USOC
4-Pair

Pair ID	PIN #
T1	5
R1	4
T2	3
R2	6
T3	2
R3	7
T4	1
R4	8

USOC
1-, 2-, or 3-Pair

Pair ID	PIN #
T1	4
R1	3
T2	2
R2	5
T3	1
R3	6

ETHERNET 10 BASE-T

Ethernet 10BASE-T wiring specifies an 8-position jack but uses only two pairs. These are pairs two and three of TIA schemes.

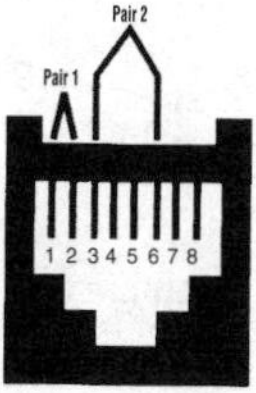

Pair ID	PIN #
T1	1
R1	2
T2	3
R2	6

WIRING DIAGRAMS

RJ31X

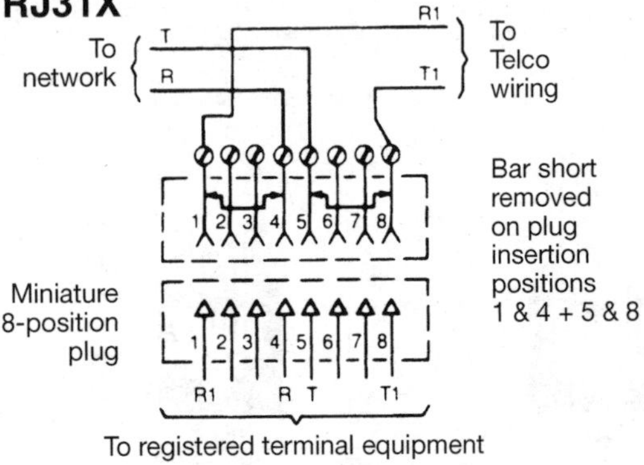

To network
T
R
R1
T1
To Telco wiring
1 2 3 4 5 6 7 8
Bar short removed on plug insertion positions 1 & 4 + 5 & 8
Miniature 8-position plug
1 2 3 4 5 6 7 8
R1 R T T1
To registered terminal equipment

RJ32X

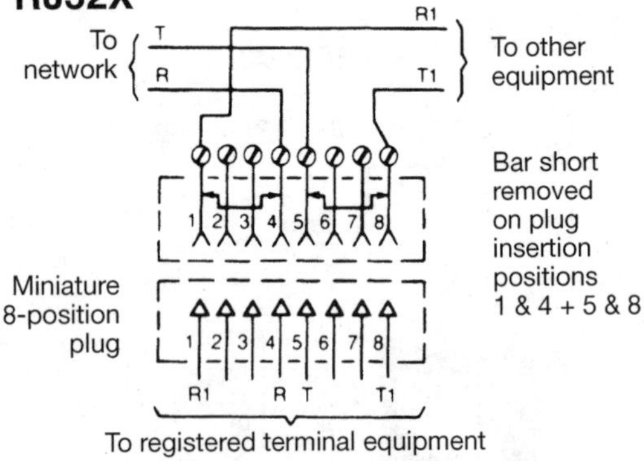

To network
T
R
R1
T1
To other equipment
1 2 3 4 5 6 7 8
Bar short removed on plug insertion positions 1 & 4 + 5 & 8
Miniature 8-position plug
1 2 3 4 5 6 7 8
R1 R T T1
To registered terminal equipment

WIRING DIAGRAMS *(cont'd)*

RJ33X

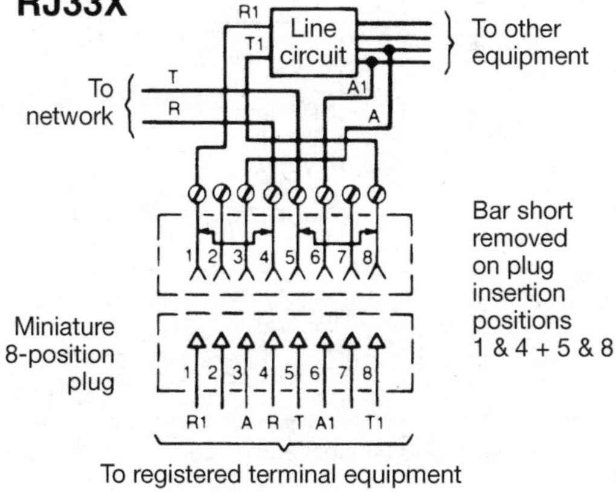

RJ34X

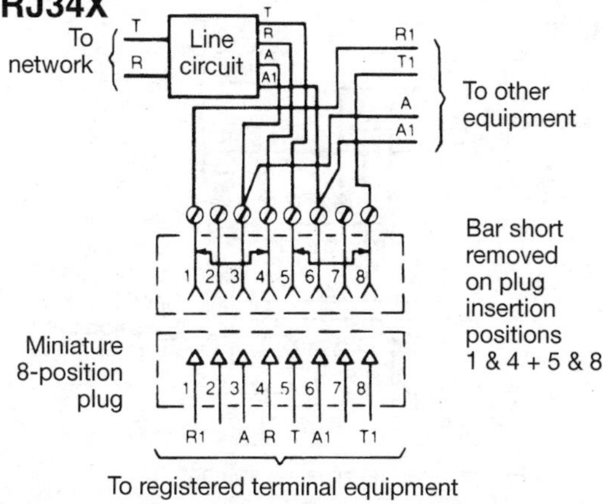

WIRING DIAGRAMS *(cont'd)*

RJ35X

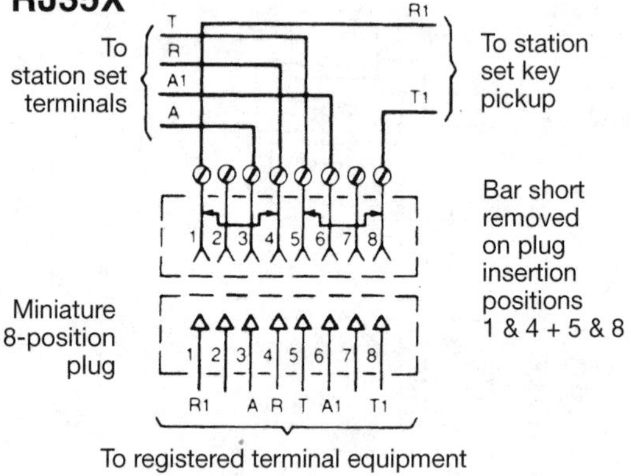

RJ36X

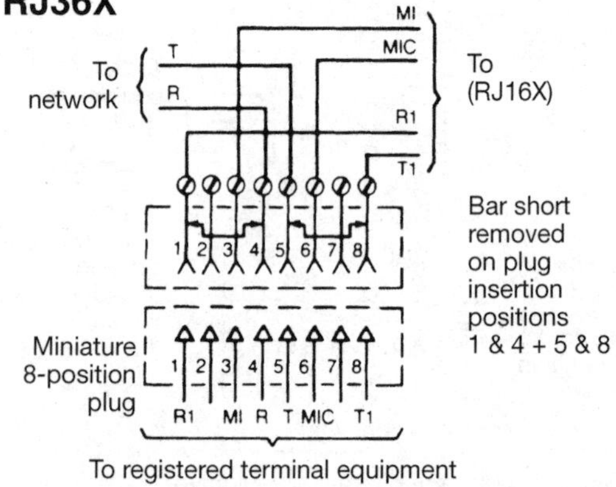

WIRING DIAGRAMS *(cont'd)*

RJ37X

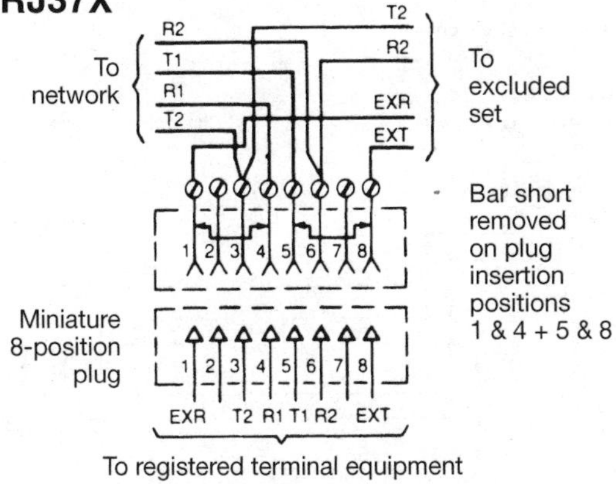

RJ38X

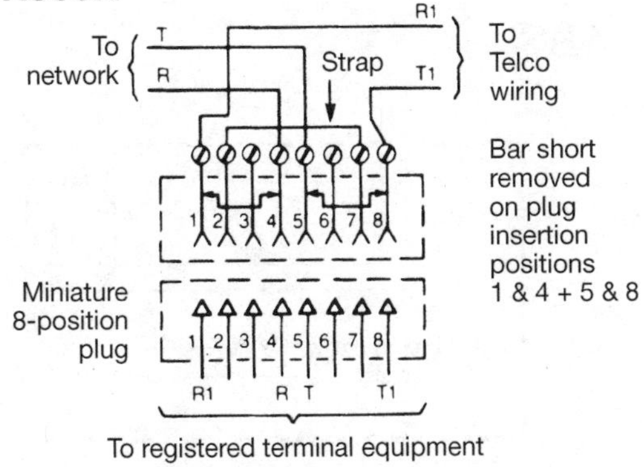

WIRING DIAGRAMS *(cont'd)*

Electrical Network Connection
Single line bridged tip and ring with programming resistor.

Mechanical Arrangement
Miniature eight-position keyed jack.

Typical Usage
Connects computers and other data equipment to the telephone network.

RJ41S

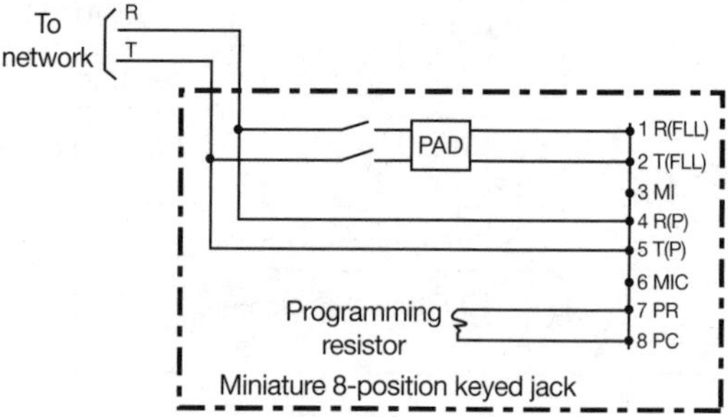

RJ45S

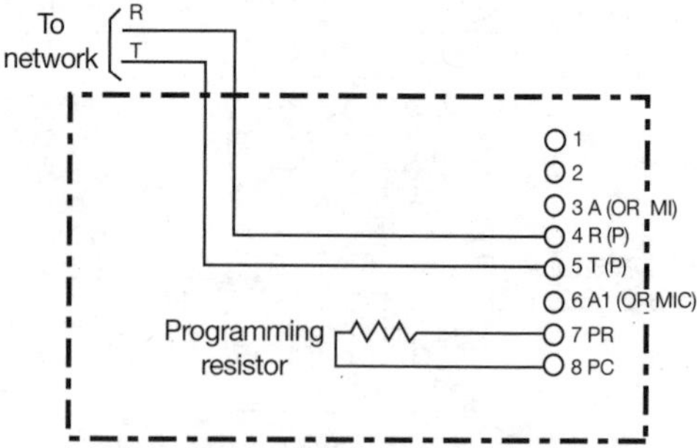

WIRING DIAGRAMS *(cont'd)*

Electrical Network Connection
Up to four line T/R.

Mechanical Arrangement
Miniature eight-position modular jack.

Typical Usage
Connects up to four lines to a single telephone set or other device. Commonly used for telephones requiring separate power pairs and/or separate signaling pairs.

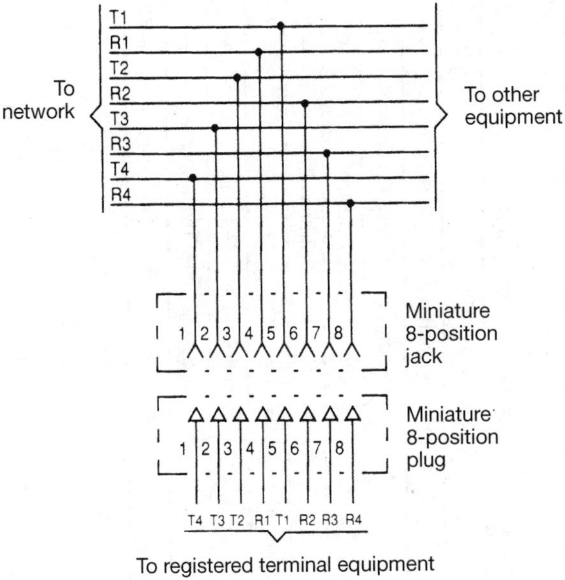

ISDN ASSIGNMENT OF CONTACT NUMBERS

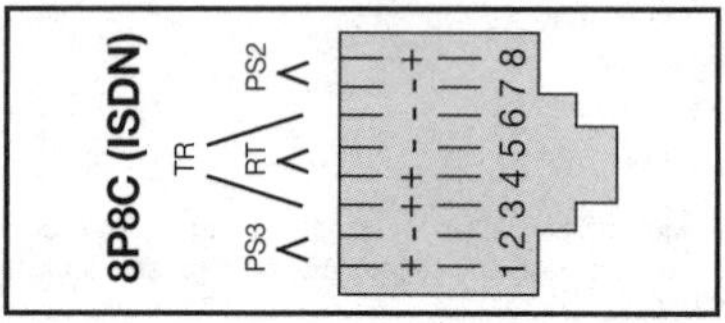

Table contact assignments for plugs and jacks:

Contact Number	TE	NT	Polarity
1	Power source 3	Power sink 3	+
2	Power source 3	Power sink 3	–
3	Transmit	Receive	+
4	Receive	Transmit	+
5	Receive	Transmit	–
6	Transmit	Receive	–
7	Power sink 2	Power source 2	–
8	Power sink 2	Power source 2	+

TYPICAL WIRING METHODS

LOOP SERIES WIRING

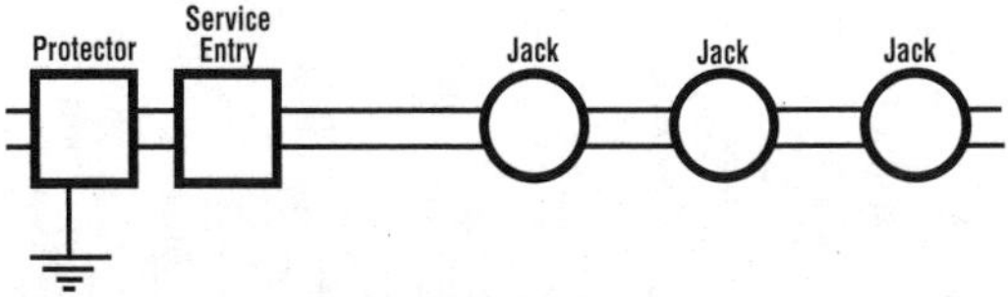

PARALLEL DISTRIBUTION WIRING

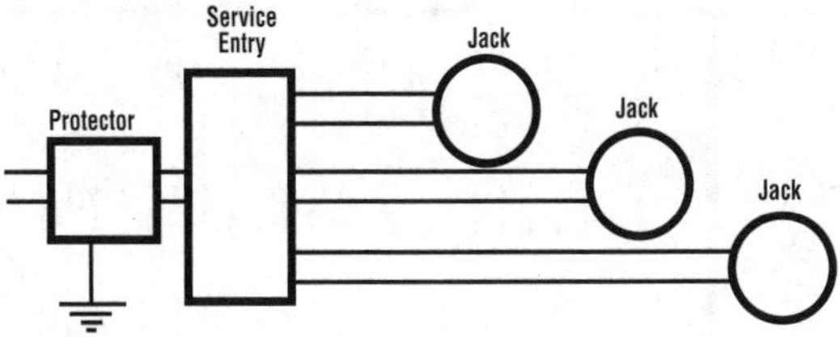

2-LINE SYSTEM

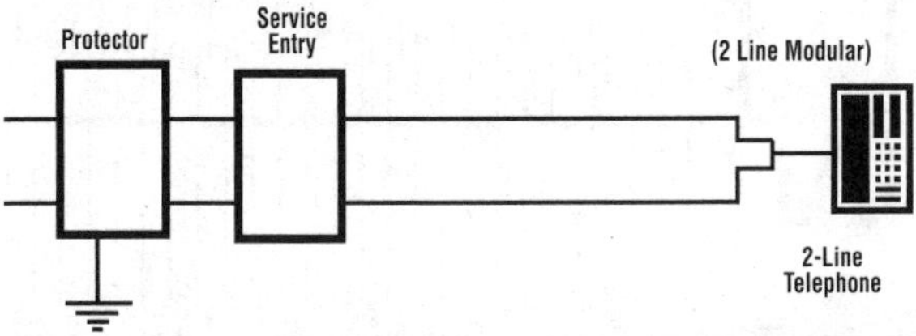

ELECTRONIC KEY SYSTEMS

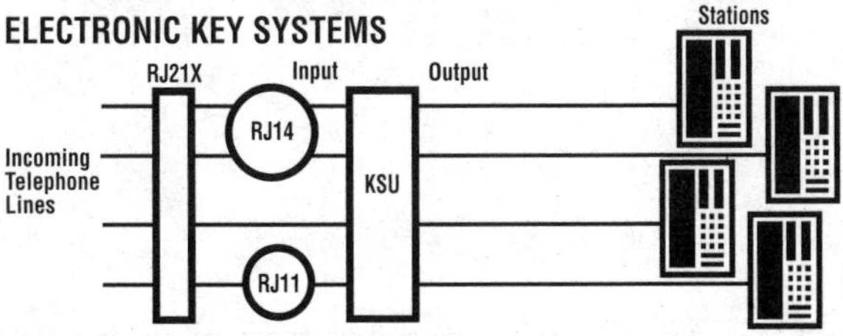

66 BLOCK WIRING & CABLE COLOR CODING

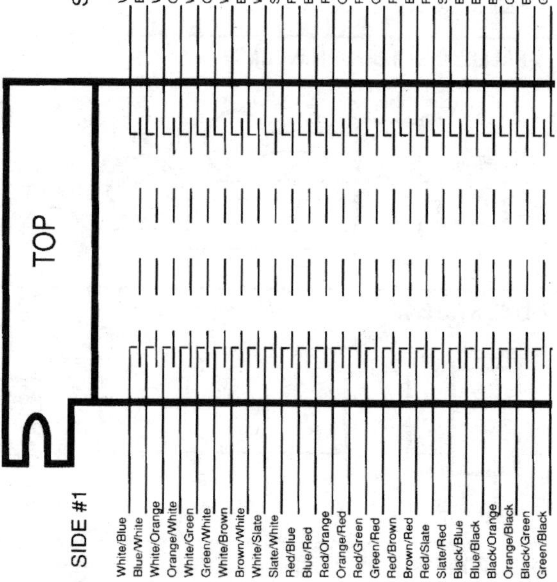

PAIR CODE			SIDE #1	SIDE #2
Pair 1	Tip	26	White/Blue	White/Blue
	Ring	1	Blue/White	Blue/White
Pair 2	Tip	27	White/Orange	White/Orange
	Ring	2	Orange/White	Orange/White
Pair 3	Tip	28	White/Green	White/Green
	Ring	3	Green/White	Green/White
Pair 4	Tip	29	White/Brown	White/Brown
	Ring	4	Brown/White	Brown/White
Pair 5	Tip	30	White/Slate	White/Slate
	Ring	5	Slate/White	Slate/White
Pair 6	Tip	31	Red/Blue	Red/Blue
	Ring	6	Blue/Red	Blue/Red
Pair 7	Tip	32	Red/Orange	Red/Orange
	Ring	7	Orange/Red	Orange/Red
Pair 8	Tip	33	Red/Green	Red/Green
	Ring	8	Green/Red	Green/Red
Pair 9	Tip	34	Red/Brown	Red/Brown
	Ring	9	Brown/Red	Brown/Red
Pair 10	Tip	35	Red/Slate	Red/Slate
	Ring	10	Slate/Red	Slate/Red
Pair 11	Tip	36	Black/Blue	Black/Blue
	Ring	11	Blue/Black	Blue/Black
Pair 12	Tip	37	Black/Orange	Black/Orange
	Ring	12	Orange/Black	Orange/Black
Pair 13	Tip	38	Black/Green	Black/Green
	Ring	13	Green/Black	Green/Black

66 BLOCK WIRING & CABLE COLOR CODING *(cont'd)*

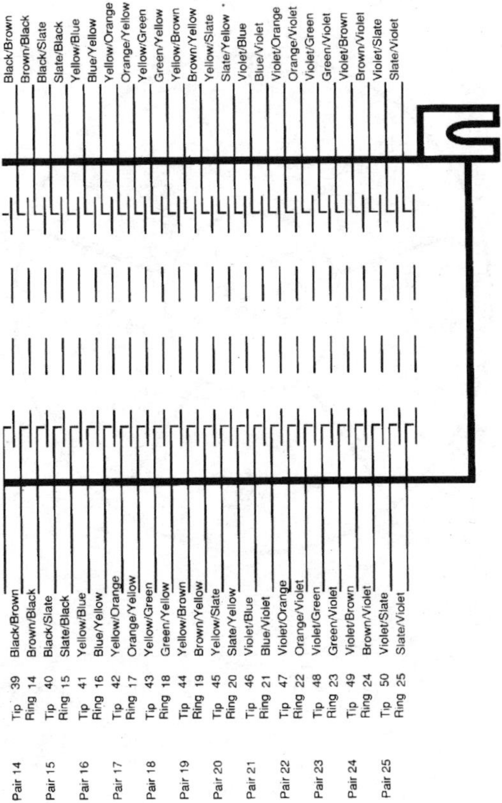
Black/Brown
Brown/Black
Black/Slate
Slate/Black
Yellow/Blue
Blue/Yellow
Yellow/Orange
Orange/Yellow
Yellow/Green
Green/Yellow
Yellow/Brown
Brown/Yellow
Yellow/Slate
Slate/Yellow
Violet/Blue
Blue/Violet
Violet/Orange
Orange/Violet
Violet/Green
Green/Violet
Violet/Brown
Brown/Violet
Violet/Slate
Slate/Violet
Pair 14 Tip 39 Black/Brown
Ring 14 Brown/Black
Pair 15 Tip 40 Black/Slate
Ring 15 Slate/Black
Pair 16 Tip 41 Yellow/Blue
Ring 16 Blue/Yellow
Pair 17 Tip 42 Yellow/Orange
Ring 17 Orange/Yellow
Pair 18 Tip 43 Yellow/Green
Ring 18 Green/Yellow
Pair 19 Tip 44 Yellow/Brown
Ring 19 Brown/Yellow
Pair 20 Tip 45 Yellow/Slate
Ring 20 Slate/Yellow
Pair 21 Tip 46 Violet/Blue
Ring 21 Blue/Violet
Pair 22 Tip 47 Violet/Orange
Ring 22 Orange/Violet
Pair 23 Tip 48 Violet/Green
Ring 23 Green/Violet
Pair 24 Tip 49 Violet/Brown
Ring 24 Brown/Violet
Pair 25 Tip 50 Violet/Slate
Ring 25 Slate/Violet

150-PAIR PIC CABLE CORE ARRANGEMENT

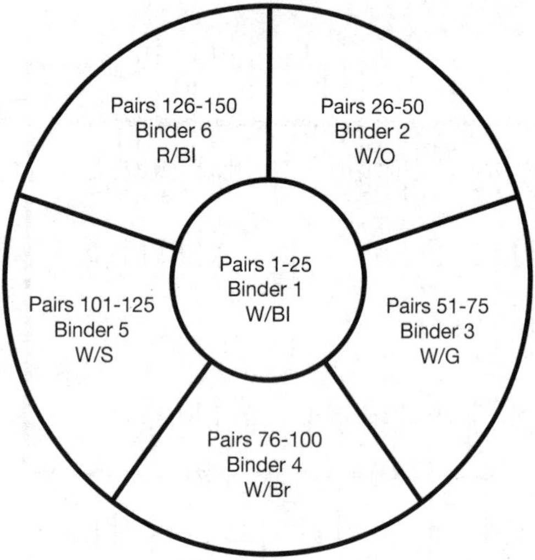

GENERAL RULES FOR PLANNING AND PERFORMING A TELEPHONE INSTALLATION

1. Determine where you want to place the modular outlets.
2. Determine which type of outlet is best for each location. If the jack is likely to be exposed to excessive dust or dirt, use jacks with protective covers.
3. Determine the best path to run the wiring from the NI or other existing telephone company-provided modular jack to each of the new outlets. Place bridges where two or more paths come together.
4. Inventory the tools you'll need to do the wiring job, such as:
   - Screwdriver with insulated handle
   - A pair of diagonal cutters with insulated grips, to cut wire
   - A tool to strip the wire coating off without damaging any of the four conductors
   - Hammer or staple gun for staples used to attach wire to wall or baseboard
   - Drill, with appropriate sized bits, to drill holes for screws, anchors and toggle bolts
   - Key hole saw, if a hole in the wall is necessary, and a drill with a large enough bit to make a hole for the saw blade
5. DO NOT place connections to wiring in outlet or junction boxes containing other electrical wiring.
6. Avoid the following if possible:
   - damp locations
   - locations not easily accessible
   - temporary structures
   - wire runs that support lighting, decorative objects, etc.
   - hot locations, such as steam pipes, furnaces, etc.
   - locations that subject wire and cable to abrasion
7. Place telephone wire at least six feet from lightning rods and associated wires, and at least six inches from other kinds of conductors (e.g., antenna wires, wires from transformers to neon signs, etc.) steam or hot water pipes and heating ducts.
8. Do not connect any external power sources to inside wire or outlets.
9. Do not run conductors between separate buildings.
10. Do not expose conductor to mechanical stress, such as being pinched when a door or window closes on it.
11. Do not place wire where it would allow a person to use the phone while in a bathtub, shower, swimming pool or other hazardous locations.
12. Do not try to pull or push wire behind walls when electric wiring is already present in the wall area.

GENERAL RULES FOR PLANNING AND PERFORMING A TELEPHONE INSTALLATION *(cont'd)*

13. Use only bridged connections if it is necessary to establish a splice of two or more wires.
14. Place connecting blocks and jacks high enough to remain moisture-free during normal floor cleaning.
15. Do not attach jacks so that the opening faces upward—this increases the potential for damage from dirt and dust.
16. Wires should run horizontally and vertically in straight lines, and should be kept as short as possible between bridges and other connections.
17. Run exposed wiring along door and window casings, baseboards, trim, and the underside of moldings, so it will not be conspicuous or unsightly.
18. Wood surfaces are better for fastening wire and attaching connecting blocks, jacks, and bridges. When attaching hardware to walls, place fasteners in studs (wooden beams behind the walls) whenever possible.
19. If drilling through walls, floors, ceilings, be careful to avoid contacts with concealed hazards, such as electrical wiring, gas pipes, steam or hot water pipes, etc.
20. If installing cables next to grating, metal grillwork, etc., use a wire guard or other protective barrier to resist abrasion.
21. Always fasten cables to cement or cinder blocks with screw anchors, drive anchors, or masonry fasteners.
22. Avoid running cables outside whenever possible. If exterior wiring is necessary, drill holes through wooden window or door frames and slope entrance holes upward from the outside. Try to use rear and side walls so the wire will not be as noticeable; place horizontal runs out of reach of children; and avoid placing wiring in front of signs, doors, windows, fire escapes, "drop wires" and across flat roofs.
23. When fastening wire to metal siding, the type of fastener used depends on the type of siding and the method used to install it. Extra caution should be used when working on mobile homes. Mobile homes should be properly grounded. Line voltages present an extreme danger when working on metal. Therefore, proper grounding is very important.

TELEPHONE CONNECTIONS

Typical Inside Wire

Type of Wire	Pair No.	Pair Color Matches	
2-pair Wire	1 2	Green Black	Red Yellow
3-pair Wire	1 2 3	White/Blue White/Orange White/Green	Blue/White Orange/White Green/White

Inside Wire Connecting Terminations

Wire Color		Wire Function	
2-pair wire	3-pair wire	Service w/o Dial Light	Service with Dial Light
Green Red Black Yellow	White/Blue Blue/White White/Orange Orange/White	Tip Ring Not Used Ground	Tip Ring Transformer Transformer

Typical Fasteners and Recommended Spacing Distances

Fasteners	Horizontal	Vertical	From Corner
Wire clamp	16 in.	16 in.	2 in.
Staples(wire)	7.5 in.	7.5 in.	2 in.
Bridle Rings*	4 ft.		2–8.5 in.*
Drive Rings**	4 ft.	8 ft.	2–8.5 in.*

*When changing direction the fasteners should be spaced to hold the wire at approximately a 45-degree angle.

**To avoid possible injury do not use drive rings below a 6 foot clearance level; instead, use bridle rings.

SEPARATION AND PHYSICAL PROTECTION FOR PREMISES WIRING

This table applies only to telephone wiring from the Network Interface or other telephone company-provided modular jacks to telephone equipment. Minimum separations between telephone wiring whether located inside or attached to the outside of buildings and other types of wiring involved are as follows. Separations apply to crossing and to parallel runs (minimum separations).

Types of Wire Involved		Minimum Separations	Wire Crossing Alternatives
Electric Supply	Bare light or power wire of any voltage	5 ft.	No Alternative
	Open wiring not over 300 volts	2 in.	See Note 1.
	Wires in conduit or in armored or nonmetallic sheath cable, or power ground wires	None	N/A
Radio & TV	Antenna lead-in and ground wires	4 in.	See Note 1.
Signal or Control Wires	Open wiring or wires in conduit or cable	None	N/A
Comm. Wires	Community Television systems coaxial cables with grounded shielding	None	N/A
Telephone Drop Wire	Using fused protectors Using fuseless protector or where no protector wiring from transformer	2 in.	See Note 1. None
Sign	Neon Signs and associated wiring from transformer		6 in. No Alternative
Lightning Systems	Lightning rods and wires		6 ft.

NOTE 1: If minimum separations cannot be obtained, additional protection of a plastic tube, wire guard, or two layers of vinyl tape extending 2 inches beyond each side of object being crossed must be provided.

TYPES OF ISDN SERVICE

BRI (Basic Rate ISDN). Consists of two 64 kbps B channels and one 16 kbps D channel for a total of 144 kbps. The basic service is intended to meet the needs of most individual users. NISDN-1 focused primarily on making basic services deployed immediately.

PRI (Primary Rate ISDN). intended for users with greater capacity requirements. Typically the channel structure is 23 B channels plus one 64 kbps D channel for a total of 1.544 Mbps. H channels can also be implemented: H0 = 384 kbps, H11 = 1536 kbps, H12 = 1920 kbps.

B-ISDN (Broadband ISDN). still in development and will support as much as 150 Mbps, but will be dependent on a complete optical fiber network. This could be a medium for future high definition television (HDTV) projects.

COMMON TYPES OF DIGITAL TELEPHONE SERVICE

Switched 56. Unlike ISDN, this service is already offered by most carriers. It creates a virtual network over existing public phone lines with a 56 Kbps data rate. This service is cheap, but slow; therefore, it is ideal for intermittent data swapping between WANs.

SMDS (Switched Multi-megabit Data Services). Using a connectionless networking plan, each SMDS packet has its own address and does not require a virtual circuit. Proposed speeds are from 1.5 Mbps to 45 Mbps using a fixed-length packet of 53 KB. Many regional carriers are beginning to offer this service for local traffic.

ATM (Asynchronous Transfer Mode). Using the same 53 KB packets as SMDS, ATM uses virtual circuits to transfer data at speeds of 34 Mbps to multiple gigabits per second. The CCITT has decided on ATM as the transport standard for broadband ISDN when it becomes available. ATM is expected to be fully supported by the phone networks in 1995 or 1996.

ISDN CABLING

ISDN Basic Rate Interface (BRI) is provided by a carrier from a central office (CO) switch to the customer premise with a two wire U-loop RJ-45 connector on the center pins 4 and 5. The connection sequence is the following:

RJ45 Plug	
1 N/C	5 U-loop network connection
2 N/C	6 N/C
3 N/C	7 N/C
4 U-loop network connection	8 N/C

ISDN NETWORK TERMINATION (NT)

The Network Termination is a Power Supply and an NT1. In North America this functionality can be provided in the terminal equipment. The connection pattern is the following:

RJ45 Plug for U + PS2	
1 N/C	5 U-loop network connection
2 N/C	6 N/C
3 N/C	7 −48 VDC
4 U-loop network connection	8 −48 VDC Return

ISDN U-LOOP

ISDN Basic Rate Interface (BRI) is provided by a carrier from a central office (CO) switch to the customer premise with a two wire U-loop RJ-45 connector on the center pins 4–5.

RJ45 PLUG	
1 N/C	5 U-loop network connection
2 N/C	6 N/C
3 N/C	7 N/C
4 U-loop network connection	8 N/C

RJ45 PLUG FOR U + PS2	
1 N/C	5 U-loop network connection
2 N/C	6 N/C
3 N/C	7 −48 VDC
4 U-loop network connection	8 −48 VDC Return

PHANTOM TELEPHONE CIRCUIT

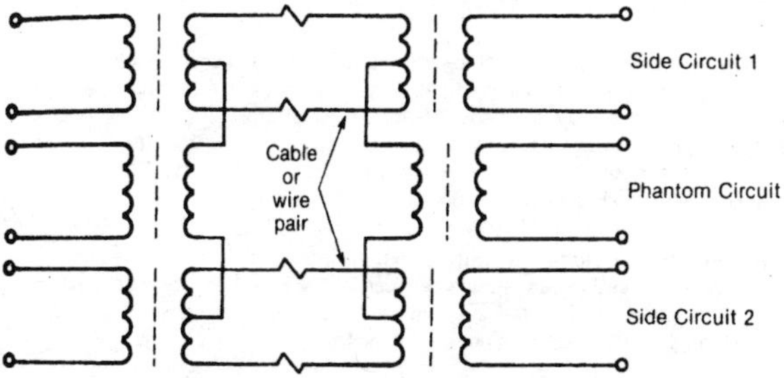

STANDARD DTMF PAD AND FREQUENCIES

(Low Group)	1209Hz	1336Hz	1477Hz	1633Hz
697Hz >	1	2	3	A
770Hz >	4	5	6	B
825Hz >	7	8	9	C
941Hz >	*	0	#	D

(High Group)

COMPUTER TELEPHONY TERMS

Following are some of the key terms that are used for defining links between data communications and telecommunications:

ADSL—Asymmetric Digital Subscriber Line. A method of carrying high-speed traffic over existing copper twisted-pair wires. Currently in the trial phase, ADSL offers three channels: a high-speed (between 1.5Mbps and 6.1Mbps) downlink from the carrier to the customer, a full-duplex data channel at 576Kbps, and a plain old telephone service (POTS) channel. A key feature of ADSL is that POTS is available even if the extra ADSL services fail.

ANI—Automatic Number Identification. A system that identifies the telephone number of the calling party for the call recipient, is known to most consumers as caller ID. When using a T-1 line, the ANI information also includes the geographic coordinates of the originating call's central office.

DS-0 A Digital Signal Level Zero. This is one of the 64,000bps circuits in a T-1 or E-1 line. It consists of 8-bit frames transmitted at 8,000 frames per second. The usable bit rate is often only 56,000bps. DS-1, or Digital Signal Level One, is often used as synonym for T-1, but it more precisely refers to the signaling and framing specifications of a T-1 line. See T-1.

DTMF—Dual-Tone Multifrequency. A description of the audio bleeps you hear when you dial a Touch-Tone telephone. Each row and each column of keys is assigned a separate frequency. Each key's frequency is produced by combining row and column frequencies. For example, the second column's assigned frequency is 1,336Hz, and the third row has 852Hz; pressing the number 7 generates both those tones. By decoding the two frequencies, the telephone company's central office, your PBX, or an interactive voice response system can detect which button was pressed. ("Touch-Tone" is AT&T's trademark for DTMF.)

Frame relay. This refers to a shared-bandwidth wide area network based on a subset of High-level Data-link Control (HDLC) called LAP-D (link access procedure-D channel). Frame relay is designed to be carried over high-speed, high-accuracy links such as T-1 or the still emerging T-3; a 56Kbps line is the most common implementation. Individual frames can vary in size, but they are usually 4,096 bytes. Users reserve a specific data rate called the CIR, or committed information rate, but users can attempt to burst data at higher rates. Extra frames are discarded if the carrier's network doesn't have sufficient capacity.

HDSL—The High bit-rate Digital Subscriber Line. This circuit offers a full-duplex 784Kbps connection over two twisted pairs. HDSL can carry either a full T-1 connection over two twisted pairs or a fractional T-1 connection over a single twisted pair of wires.

COMPUTER TELEPHONY TERMS *(cont'd)*

ISDN Integrated Services Digital Network. A telephone system service that provides access to both the public-switched telephone network and to packet-switched services (such as X.25 and frame relay). ISDN offers two types of channels: B (bearer), which are 64Kbps voice channels, and D (delta), which are channels for setup, coordination, and control. Telephone companies offer ISDN in two main varieties: basic rate interface (BRI), which contains two B channels and an 8Kbps D channel, and primary rate interface (PRI), which has 23 B channels and a 64kbps D channel.

IVR—Interactive Voice Response. The basic voice-mail system that can decode DTMF signals, which is how your call can be routed without the aid of a real human being.

PBX—Private Branch Exchange. A telephone switch used within a business or other enterprise—as opposed to the switches used at a public telecommunication service provider's central office (CO). PBXs might offer basic telephone service, some level of computer-telephony integration (CTI), voice mail, or other features. When you dial within your company, the PBX provides the dial tone; when you dial 8 or 9 for an outside line, the CO provides the dial tone.

T-1 T-1 is a North American standard for point-to-point digital circuits over two twisted pairs. A T-1 line carries 24 64,000bps channels (also known as DS-0) for a total usable bit rate of 1,536,000bps (if you include extra bits used to synchronize the frames, the actual bandwidth is 1,544,000bps). Customers may lease a fractional T-1, using only some of the 24 T-1 slots. A T-1C contains two T-1 lines; T-2 supports four T-1 circuits. A T-3 communications circuit supports 28 T-1 circuits, and a T-4 consists of 168 T-1 circuits. E-1 through E-5 are similar standards used in Europe and Japan, but they offer different numbers of channels. See DS-0.

TAPI Telephony Application Programming Interface is a method promoted by Microsoft and Intel for letting PCs control telephones. TAPI applications can dial a telephone from within software and check caller ID, as well as perform other functions. TAPI links PC workstations and individual telephones, as opposed to TSAPI (see below), which links the PC server to the PBX.

TSAPI—Telephony Services Application Programming Interface. A system promoted by Novell and AT&T to integrate PCs and PBX servers. TSAPI allows computer control of most aspects of the local telephone system. Contrast this to TAPI, which links the PC and the local telephone sitting on the user's desk.

X.25 This term refers to a standard for packet-based wide area networks. For both leased lines, such as T-1s, and public-switched connections, like ISDN's B-channel links, you pay by the minute or month. However, an X.25 connection has the advantage of being measured and billed by the number of packets or bytes actually sent or received.

TELECOMMUNICATIONS ACRONYMS

AC	Alternating Current
ACD	Automatic Call Distributor
ACTS	Automatic Coin Telephone System
ADPCM	Adaptive Differential Pulse Code Modulation
AF	Audio Frequency
AIS	Automatic Intercept System
ALI	Automatic Location Information
AMA	Automatic Message Accounting
AMI	Alternate Mark Inversion
ANI	Automatic Number Identification
AOS	Alternate Operator Service
APD	Avalanche Photo Diode
ARQ	Automatic Repeat Request
ASCII	American Standard Code for Information Interexchange
ATM	Asynchronous Transfer Mode
AWG	American Wire Gauge
B8ZS	Bipolar with 8-Zero Substitution
BCD	Blocked Calls Delayed
BCH	Blocked Calls Held
BCR	Blocked Calls Released
BER	Bit Error Rate
BHCA	Busy Hour Call Attempts
BIOS	Basic Input/Output System
BLER	Block Error Rate
BOC	Bell Operating Company
BRI	Basic Rate Interface
BSA	Basic Service Arrangement
BSC	Binary Synchronous Communications
BSE	Basic Service Element
CAC	Carrier Access Code
CAD	Computer Aided Dispatch; Computer Aided Design
CAS	Centralized Attendant Service
CATV	Community Antenna Television
CBX	Computer Branch Exchange
CCIS	Common Channel Interoffice Signaling
CCITT	Consultative Committee on International Telephone and Telegraph
CCS	Centum Call Seconds
CCS	Common Channel Signaling
CCTV	Closed Circuit Television
CDO	Community Dial Office
CDR	Call Detail Recorder
CELP	Code Excited Linear Prediction
CEPT	Conference European on Post and Telecommunications
CGSA	Cellular Geographic Serving Area
CLASS	Custom Local Area Signaling Services

TELECOMMUNICATIONS ACRONYMS *(cont'd)*

CSN	Complementary Network Services
CO	Central Office
CODEC	Coder/Decoder
CPE	Customer Premises Equipment
CPU	Central Processing Unit
CRC	Cyclical Redundancy Checking
CRSO	Cellular Radio Switching Office
CSM/CD	Carrier Sense Multiple Access with Collision Detection
CSU	Channel Service Unit
DAMA	Demand Assigned Multiple Access
dB	Decibel
DBS	Direct Broadcast Satellite
DC	Direct Current
DCE	Data Circuit-Terminating Equipment
DCS	Digital Crossconnect System
DDD	Direct Distance Dialing
DID	Direct Inward Dialing
DMI	Digital Multiplexed Interface
DNIC	Data Network Identification Code
DNIS	Dialed Number Identification Service
DOC	Dynamic Overload Control
DOV	Data Over Voice
DQDB	Distributed Queue Dual Bus
DSI	Digital Speech Interpolation
DSX	Digital Service Crossconnect
DTE	Data Terminal Equipment
DTMF	Dual Tone Multifrequency
EAS	Extended Area Service
EBCDIC	Expanded Binary Coded Decimal Interexchange Code
EDI	Electronic Data Interchange
EFS	Error Free Seconds
EIRP	Effective Isotropic Radiated Power
EMI	Electromagnetic Interference
ERL	Echo Return Loss
ESF	Extended Super Frame
ESP	Enhanced Service Provider
ESS	Electronic Switching System
ETN	Electronic Tandem Network
FDDI	Fiber Distributed Data Interface
FDMA	Frequency Division Multiple Access
FEC	Forward Error Correction
FX	Foreign Exchange
gb/s	Gigabits Per Second
GHz	Gigahertz
GOS	Grade of Service

TELECOMMUNICATIONS ACRONYMS *(cont'd)*

HDTV	High Definition Television
Hz	Hertz
IC	Independent Company
IDF	Intermediate Distributing Frame
IEC	Interexchange Carrier
IMTS	Improved Mobile Telephone Service
INMARSAT	International Maritime Satellite Service
INWATS	Inward Wide Area Telephone Service
ISDN	Integrated Services Digital Network
IVR	Interactive Voice Response
kb/s	Kilobits Per Second
KHz	Kilohertz
KTS	Key Telephone System
LAN	Local Area Network
LAPD	Link Access Procedure D
LATA	Local Access Transport Area
LCR	Least Cost Routing
LEC	Local Exchange Carrier
LED	Light Emitting Diode
LIT	Line Insulation Test
LLC	Logical Link Control
LMX	L Multiplex
LTD	Local Test Desk
MAP/TOP	Manufacturing Automation Protocol/Technical and Office Protocol
mb/s	Megabits Per Second
MCVD	Modified Chemical Vapor Deposit
MDF	Main Distributing Frame
MHz	Megahertz
MLN	Main Listed Number
MTBF	Mean Time Between Failures
MTS	Message Telephone Service
MTTR	Mean Time to Repair
NAU	Network Access Unit
NCTE	Network Channel Terminating Equipment
NID	Network Interface Device
NMCC	Network Management Control Center
NPA	Numbering Plan Area
NT1	Network Termination 1
NT12	Network Termination 12
NT2	Network Termination 2
NTN	Network Terminal Number
ONA	Open Network Architecture
OPX	Off-Premise Extension
OSI	Open Systems Interconnect
OSS	Operations Support System

TELECOMMUNICATIONS ACRONYMS *(cont'd)*

P/AR	Peak-to-Average Ratio
PABX	Private Automatic Branch Exchange
PAD	Packet Assembler/Disassembler
PAM	Pulse Amplitude Modulation
PASP	Public Service Answering Point
PBX	Private Branch Exchange
PCM	Pulse Code Modulation
PIC	Primary Interexchange Carrier
PIN	Personal Identification Number
Pixel	Picture Element
POP	Point of Presence
PRI	Primary Rate Interface
PSK	Phase Shift Keying
PSTN	Public Switched Telephone Network
QAM	Quadrature Amplitude Modulation
RCF	Remote Call Forwarding
RMATS	Remote Maintenance and Testing System
rms	Root Mean Square
rn	Reference Noise
ROTL	Remote Office Test Line
RSL	Received Signal Level
SCC	Satellite Communications Control
SDN	Software Defined Network
SF	Single Frequency
SMDR	Station Message Detail Recording
SMDS	Switched Multimegabit Data Service
SMSA	Standard Metropolitan Statistical Area
SNA	Systems Network Architecture
SONET	Synchronous Optical Network
SPC	Stored Program Control
SRL	Singing Return Loss
SSB	Single Sideband
SSTDMA	Spacecraft Switched Time Division Multiple Access
TCM	Trellis-Coded Modulation
TDM	Time Division Multiplexing
TDMA	Time Division Multiple Access
TDR	Time Domain Reflectometer
TE1	Terminal Equipment Type 1
TE2	Terminal Equipment Type 2
TLP	Transmission Level Point
TUR	Traffic Usage Recorder
UCD	Uniform Call Distribution
UPS	Uninterruptable Power Supply
VF	Voice Frequency
VNL	Via Net Loss

CABLE MODEMS

The term *Cable Modem* is quite new and refers to a modem that operates over the ordinary cable TV network cables to provide broadband Internet services. The Cable Modem is connected to a TV outlet, and the cable TV system connects a Cable Modem Termination System (CMTS) at the Head-End of the network. Actually the term "Cable Modem" is a bit misleading, as a Cable Modem works more like a Local Area Network (LAN) interface than as a modem.

CABLE MODEM OPERATION

- The signal transfer rate for a cable modem is typically between 3 and 50 Mbit/s, and the distance can be up to 100 km or more.
- The Cable Modem Termination System (CMTS) can talk to all the Cable Modems (CM's), but the Cable Modems can only talk to the CMTS. If two Cable Modems need to talk to each other, the CMTS will have to relay the messages.
- A cable modem will work with most operating systems and hardware platforms, including Mac, UNIX, laptop computers etc.
- Another interface for external Cable Modems is USB, which has the advantage of installing much faster (something that matters, because the cable operators are normally sending technicians out to install each and every Cable Modem.) The downside is that you can only connect one PC to a USB based Cable Modem.

CABLE MODEM INSTALLATION CONCERNS

- When installing a Cable Modem, a power splitter and a new cable is usually required. The splitter divides the signal for the "old" installations and the new segment that connects the Cable Modem. No TV-sets are accepted on the new string that goes to the Cable Modem.
- The transmitted signal from a Cable Modem can be so strong that any TV sets connected on the same string can be disturbed . The isolation of the splitter may not be sufficient, and an extra high-pass filter may be required in the string that goes to the TV sets. The high-pass filter allows only the TV-channel frequencies to pass, and blocks the upstream frequency band. The other reason for the filter is to block ingress in the low upstream frequency range from the in-house wiring. Noise injected at each individual residence accumulates in the upstream path towards the head-end, so it is essential to keep it at a minimum at every single residence that needs Cable Modem service.

BROADBAND TERMS

Here is a short list of some of the other technical terms and acronyms that you may stumble across in trying to understand the cable modem world.

CATV: Cable TV system. Can be all coaxial or HFC (Hybrid Fiber Coax) based.

Cable Modem (CM): Client device for providing data over a cable TV network.

Channel: A specific frequency and bandwidth combination. Used in this context as TV channels for television services and downstream data for cable modems.

CMTS: Cable Modem Termination System. Central device for connecting the cable TV network to a data network like the Internet. Normally placed in the headend of the cable TV system.

CPE: Customer Premises Equipment. Used to describe the PC and/or other equipment, that the customer may want to connect to the cable modem.

DHCP: Dynamic Host Configuration Protocol. This protocol provides a mechanism for allocating IP (Internet Protocol) addresses dynamically so that addresses can be reused. Often used for managing the IP addresses of all the cable modems in a cable plant and the PC's connected to the cable modems.

DOCSIS: Data Over Cable Service Interface Specification. The dominating cable modem standard. Defines technical specifications for both cable modem and CMTS.

Downstream: The data flowing from the CMTS to the cable modem.

Downstream Frequency: The frequency used for transmitting data from the CMTS to the cable modem. Normally in the 42/65-850 MHz range depending on the actual cable plant capabilities.

Headend: Central distribution point for a CATV system. Video signals are received here from satellites and maybe other sources, frequency converted to the appropriate channels, combined with locally originated signals, and rebroadcast onto the HFC plant. The headend is where the CMTS is normally located.

HFC: Hybrid Fiber-Coaxial (cable network). Older CATV systems were provisioned using only coaxial cable. Modern systems use fiber transport from the headend to an optical node located in the neighborhood to reduce system noise. Coaxial cable runs from the node to the subscriber. The fiber plant is generally a star configuration with all optical node fibers terminating at a headend. The coaxial cable part of the system is generally a trunk-and-branch configuration.

MAC Layer: Media Access Control sublayer in the network stack.

MCNS: Multimedia Cable Network System Partners Ltd. The consortium behind the DOCSIS standard for cable modems.

Minislot: Basic timeslot unit used for upstream data bursts in the DOCSIS standard.

BROADBAND TERMS *(cont'd)*

MSO: Multiple Service Operator. A cable TV service provider that also provides other services such as data and/or voice telephony.

QAM: Quadrature Amplitude Modulation. A method of modulating digital signals using both amplitude and phase coding. Used for downstream and can be used for upstream.

QPSK: Quadrature Phase-Shift Keying. A method of modulating digital signals using four phase states to code two digital bits per phase shift.

Ranging: The process of automatically adjusting transmit levels and time offsets of individual modems, in order to make sure the bursts coming from different modems line up in the right timeslots and are received at the same power level at the CMTS.

SID (Service ID): Used in the DOCSIS standard to define a particular mapping between a cable modem (CM) and the CMTS. The SID is used for the purpose of upstream bandwidth allocation and class-of-service management.

Subscriber Unit (SU): An alternate term for cable modem.

Upstream: The data flowing from the CM to the CMTS.

Upstream Frequency: The frequency used to transmit data from the CM to the CMTS. Normally in the 5-42 MHz range for US systems and 5-65 MHz for European systems.

BASIC ADSL CONNECTION

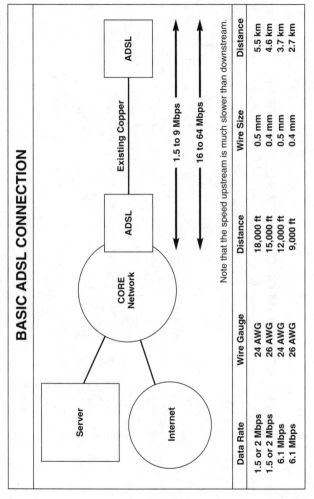

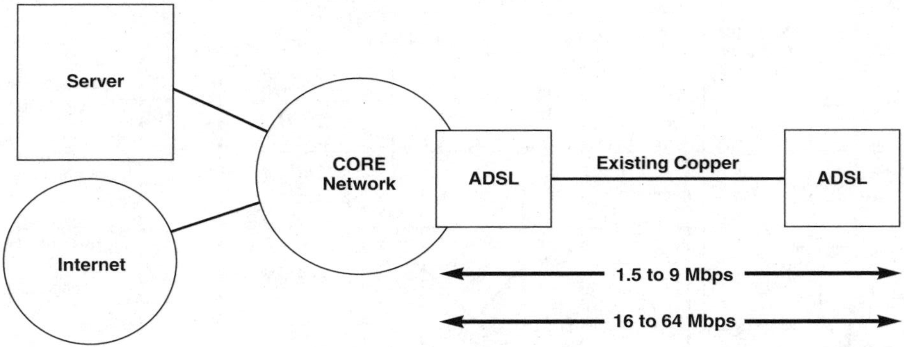

Note that the speed upstream is much slower than downstream.

Data Rate	Wire Gauge	Distance	Wire Size	Distance
1.5 or 2 Mbps	24 AWG	18,000 ft	0.5 mm	5.5 km
1.5 or 2 Mbps	26 AWG	15,000 ft	0.4 mm	4.6 km
6.1 Mbps	24 AWG	12,000 ft	0.5 mm	3.7 km
6.1 Mbps	26 AWG	9,000 ft	0.4 mm	2.7 km

BLOCK DIAGRAM OF COMPLEX ADSL NETWORK

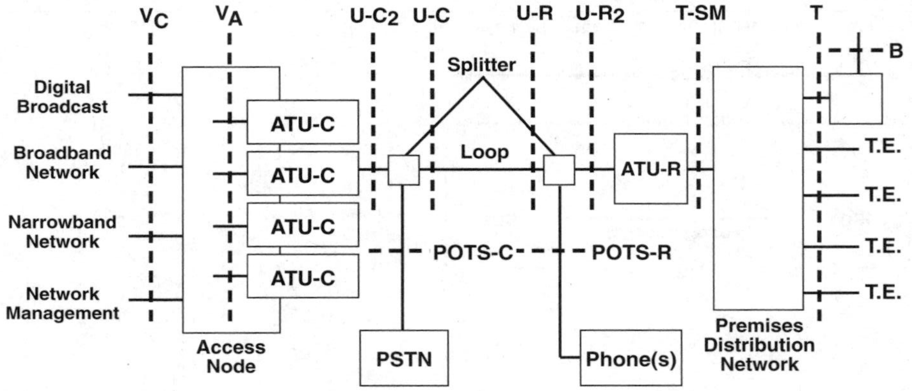

CHAPTER 7
VIDEO

BASIC TELEVISION CHANNEL

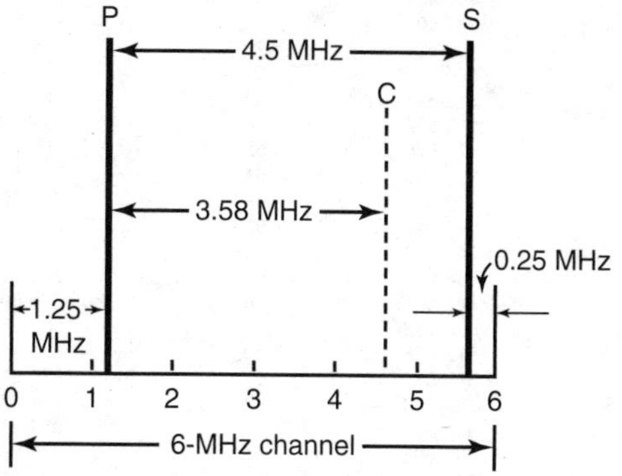

P is picture carrier; S is sound carrier,
C is color subcarrier.

TELEVISION CHANNEL ALLOCATIONS

CHANNEL NUMBER	FREQUENCY BAND, MHz	CHANNEL NUMBER	FREQUENCY BAND, MHZ
1*	—	4	66–72
2	54–60	5	76–82
3	60–66	6	82–88

TELEVISION CHANNEL ALLOCATIONS *(cont'd)*

CHANNEL NUMBER	FREQUENCY BAND, MHz	CHANNEL NUMBER	FREQUENCY BAND, MHz
7	174-180	45	656-662
8	180-186	46	662-668
9	186-192	47	668-674
10	192-198	48	674-680
11	198-204	49	680-686
12	204-210	50	686-692
13	210-216	51	692-698
		52	698-704
14	470-476	53	704-710
15	476-482	54	710-716
16	482-488	55	716-722
17	488-494	56	722-728
18	494-500	57	728-734
19	500-506	58	734-740
20	506-512	59	740-746
21	512-518	60	746-752
22	518-524	61	752-758
23	524-530	62	758-764
24	530-536	63	764-770
25	536-542	64	770-776
26	542-548	65	776-782
27	548-554	66	782-788
28	554-560	67	788-794
29	560-566	68	794-800
30	566-572	69	800-806
31	572-578	70	806-812
32	578-584	71	812-818
33	584-590	72	818-824
34	590-596	73	824-830
35	596-602	74	830-836
36	602-608	75	836-842
37	608-614	76	842-848
38	614-620	77	848-854
39	620-626	78	854-860
40	626-632	79	860-866
41	632-638	80	866-872
42	638-644	81	872-878
43	644-650	82	878-884
44	650-656	83	884-890

*The 44- to 50-MHz bank was television channel I but is now assigned to other services.

†Channels 70 to 83 are also allocated for land mobile radio. For television, these UHF channels are used for special services.

‡Channel 37 is not available for TV assignment.

CABLE TV CHANNELS

LETTER DESIGNATION	NUMBER	VIDEO CARRIER, MHz
Midband channels		
A	14	121.25
B	15	127.25
C	16	133.25
D	17	139.25
E	18	145.25
F	19	151.25
G	20	157.25
H	21	163.25
I	22	169.25
Superband channels		
J	23	217.25
K	24	223.25
L	25	229.25
M	26	235.25
N	27	241.25
O	28	247.25
P	29	253.25
Q	30	259.25
R	31	265.25
S	32	271.25
T	33	277.25
U	34	283.25
V	35	289.25
W	36	295.25
X	37	301.25
Y	38	307.25
Z	39	313.25

NUMBER	VIDEO CARRIER, MHz
Superband channels without letters	
40	319.25
41	325.25
42	331.25
43	337.25
44	343.25
45	349.25
46	355.25
47	361.25
48	367.25
49	373.25
50	379.25
51	385.25
52	391.25
53	397.25
Additional midband assignments	
54	89.25
55	95.25
56	101.25
57	107.25
58	97.25
59	103.25
Nominal channel numbers for use with digital readout converters	
A-2 or 00	109.25
A-1 or 01	115.25

HARMONICALLY RELATED CARRIERS FOR CABLE TV CHANNELS

CHANNEL NUMBER	VIDEO CARRIER, MHz	CHANNEL NUMBER	VIDEO CARRIER, MHz
00	108.00	30	258.00
01	114.00	31	264.00
02	54.00	32	270.00
03	60.00	33	276.00
04	66.00	34	282.00
05	78.00	35	288.00
06	84.00	36	294.00
07	174.00	37	300.00
08	180.00	38	306.00
09	186.00	39	312.00
10	192.00	40	318.00
11	198.00	41	324.00
12	204.00	42	330.00
13	210.00	43	336.00
14	120.00	44	342.00
15	126.00	45	348.00
16	132.00	46	354.00
17	138.00	47	360.00
18	144.00	48	366.00
19	150.00	49	372.00
20	156.00	50	378.00
21	162.00	51	384.00
22	168.00	52	390.00
23	216.00	53	396.00
24	222.00	54	72
25	228.00	55	90
26	234.00	56	96
27	240.00	57	102
28	246.00	58	402
29	252.00	59	408

PRINCIPAL TELEVISION SYSTEMS

	NTSC NORTH AND SOUTH AMERICA;* INCLUDES U.S., CANADA, MEXICO, AND JAPAN	PAL WESTERN EUROPE; INCLUDES GERMANY, ITALY, AND SPAIN	ENGLAND†	FRANCE‡	U.S.S.R.
Lines per frame	525	625	625	625	625
Frames per second	30	25	25	25	25
Field frequency, Hz	60	50	50	50	50
Line frequency, Hz	15,750	15,625	15,625	15,625	15,625
Video bandwidth, MHz	4.2	5 or 6	5.5	6	6
Channel width, MHz	6	7 or 8	8	8	8
Video modulation	Negative	Negative	Negative	Positive	Negative
Sound signal	FM	FM	FM	AM	FM
Color system	NTSC	PAL	PAL	SECAM	SECAM
Color subcarrier, MHz	3.58	4.43	4.43	4.43	4.43

*Exceptions are Argentina and Brazil, which use PAL.
†England also uses 405-line system in 5-MHz channel.
‡France also uses 819-line system in 14-MHz channel.
Note: NTSC is National Television Systems Committee; PAL is phase alteration by line: and SECAM is sequential chrominance and memory.

RADIO FREQUENCY BANDS

The main groups of radio frequencies and their wavelengths are as follows:

VLF = Very low frequencies, 3 to 30 kHz; wavelengths 100 to 10 km

LF = Low frequencies, 30 to 300 kHz; wavelengths 10 to 1 km

MF = Medium frequencies, 0.3 to 3 MHz; wavelengths 1000 to 100 m

HF = High frequencies, 3 to 30 MHz; wavelengths 100 to 10 m

VHF = Very high frequencies, 30 to 300 MHz; wavelengths 10 to 1 m

UHF = Ultrahigh frequencies, 300 to 3000 MHz; wavelengths 1000 to 100 mm

SHF = Superhigh frequencies, 3 to 30 GHz; wavelengths 100 to 10 mm

EHF = Extra-high frequencies, 30 to 300 GHz; wavelengths 10 to 1 mm

Note that wavelengths are shorter for higher frequencies. Also 1 GHz = 1×10^9 Hz = 1000 MHz.

Microwaves have wavelengths of 1 m down to 1 mm (300 GHz). The spectrum of light rays starts at frequencies of 300 GHz and up, with infrared radiation having wavelengths from 1 mm to 10 μm.

RADIO SERVICES AND TELEVISION INTERFERENCE

Principal services that can cause interference in television receivers are in the following list. The interference is generally at a harmonic frequency or image frequency. All frequencies are in megahertz.

AMATEUR RADIO These bands include 1.8 to 2, 3.5 to 4, 7 to 7.3,14 to 14.25, 21 to 21.45, 28 to 29.7, 50 to 54, 144 to 148, 220 to 225, 420 to 450.

INDUSTRIAL, SCIENTIFIC, AND MEDICAL This band is 13.36 to 14, with diathermy equipment at 13.56. The old frequencies were 27.12 and 40.66 to 40.70.

CITIZEN'S RADIO This band includes channels 1 to 40 in the band of 26.965 to 27.405 MHz for class D service.

FM BROADCAST This band is 88 to 108, with 100 channels spaced 0.2 MHz apart.

PUBLIC SAFETY (POLICE, FIRE, ETC.) These frequencies are around 33, 37, 39, 42, 44, 154, 156, and 158.

FCC FREQUENCY ALLOCATIONS FROM 30KHZ TO 300,000 MHZ

BAND	ALLOCATION	REMARKS
30-535 kHz	Includes maritime communications and navigation	500 kHz is international distress frequency
535-1605 kHz	Standard radio broadcast band	AM broadcasting
1605 kHz-30 MHz	Includes amateur radio and international short-wave broadcast	Amateur bands 3.5-4.0 MHz and 28-29.7 MHz
30-50 MHz	Government and nongovernment, fixed and mobile	Includes police, fire, forestry, highway, and railroad services
50-54 MHz	Amateur radio	6-m band
54-72 MHz	Television broadcast channels 2 to 4	Also fixed and mobile services
72-76 MHz	Government and nongovernment services	Aeronautical marker beacon on 75 MHz
76-88 MHz	Television broadcast channels 5 and 6	Also fixed and mobile services
88-108 MHz	FM broadcast	Also available for facsimile broadcast; 88—92-MHz educational FM broadcast

FCC FREQUENCY ALLOCATIONS FROM 30KHZ TO 300,000 MHZ *(cont'd)*

BAND	ALLOCATION	REMARKS
108-122 MHz	Aeronautical navigation	Localizers, radio range, and air traffic control
122-174 MHz	Government and nongovernment, fixed and mobile, amateur	144—148-MHz amateur band
174-216 MHz	Television broadcast channels 7 to 13	Also fixed and mobile services
216-470 MHz	Amateur, government and nongovernment, fixed and mobile, aeronautical navigation	Radio altimeter, glide path, and meteorological equipment; civil aviation 225-400 MHz
470-890 MHz	Television broadcasting	UHF television broadcast channels 14 to 83; translator stations in channels 70 to 83
890-3000 MHz	Aeronautical radio navigation, amateur, studio-transmitter relay, government and nongovernment, fixed and mobile	Radar bands 1300-1600 MHz; educational television 2500-2690 MHz; microwave ovens at 2450 MHz
3000-30,000 MHz	Government and nongovernment fixed and mobile, amateur, radio navigation	Superhigh frequencies (SHF); radio relay; Intelsat satellites
30,000-300,000 MHz	Experimental, government, amateur	Extremely high frequencies (EHF)

COAXIAL CABLE CONSTRUCTION

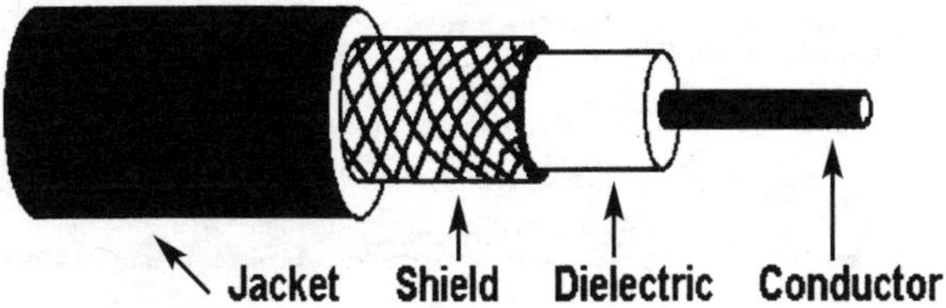

COAXIAL CABLE CLASSIFICATIONS

The NEC goes into great detail on designated cable types. Many of these requirements apply more to the cable manufacturer than to the installer. None the less, the proper cable type must be used for the installation. The NEC designations and their uses are as follow:

- **Type CATVP.** CATVP is plenum cable (hence the "P" designation), and may be used in plenums, ducts, or other spaces for environmental air.

- **Type CATVR.** CATVR is riser cable, and is made suitable (extremely fire resistant) to run installations in shafts or from floor to floor in buildings.

- **Type CATV.** CATV is general use cable. It can be used in almost any location, except for risers and plenums.

- **Type CATVX.** CATVX is a limited use cable, and is allowed only in dwellings and in raceways.

- Many coax cables are multiple-rated. In other words, their jacket is tested and suitable for several different applications. In such cases, they will be stamped with all of the applicable markings, such as CATV or CATVR.

- All coaxial cable types start with "CATV". Although you may think that the term refers to Cable Television, it does not. It refers to Community Antenna Television.

- Trade designations generally refer to the cable's electrical characteristics; specifically, the impedance of the cable. This is why different cable types (RG59U, RG58U, etc.) should not be mixed, even though they appear to be virtually identical—they have differing levels of impedance, and mixing them may degrade system performance.

SUBSTITUTIONS

- The National Electrical Code® defines a hierarchy for cable substitutions. The highest of the cable types is plenum cable—CATVP; it can be used anywhere at all. The next highest is riser cable—CATVR; it can be used anywhere at all, except in plenums. Third on the list is CATV, which can be used anywhere except in plenums or risers. Last is CATVX, which can be used only in dwellings and in raceways.
- The NEC® also permits multipurpose cables to be used for CATV work. But, as always, only plenum types of cables can be used in plenums, riser types in risers, and so on.

ATTENUATION OF TRUNK AND DISTRIBUTION CABLES (DB/100 FT AT 68°F)

FREQUENCY, MHz	T4412	T4500	T4625	T4750	T4875	T41000*
5	0.19	0.16	0.13	0.11	0.09	0.08
50	0.62	0.51	0.41	0.34	0.30	0.27
216	1.31	1.08	0.88	0.74	0.64	0.59
240	1.39	1.14	0.93	0.78	0.68	0.62
260	1.45	1.19	0.97	0.82	0.71	0.65
270	1.48	1.22	0.99	0.83	0.73	0.66'
300	1.56	1.29	1.05	0.88	0.77	0.70
325	1.63	1.34	1.09	0.92	0.80	0.74
350	1.69	1.40	1.14	0.96	0.84	0.77
375	1.75	1.45	1.18	1.00	0.87	0.80
400	1.81	1.50	1.22	1.03	0.90	0.83
450	1.93	1.59	1.30	1.10	0.96	0.89
500	2.04	1.69	1.38	1.17	1.02	0.94

* T41 000 is 1 -in. trunk cable.
Courtesy Times Fiber Communications Inc.

CCTV DESIGN

The seven concerns of CCTV (Closed-Circuit Television) design are the following:

1. **Scene lighting.** You must have at least 2 footcandles for black and white, and 5 for color work. (Except for special low-light cameras.)
2. **Contrast.** Too much contrast in the scene creates glaring reflections.
3. **Connections.** Bad connections or wrong types of wire cause an image with very low contrast. A common cause of this problem is mixing RG-58U and RG-59U cables, which look similar, but have different levels of impedance.
4. **Power voltage drops.** When cameras are located in remote locations, long runs of power wiring are required, and voltage drops may result. If they do, a loss in contrast will result.
5. **Multiple ground points.** If the image circuit is grounded at more than one point, ground loops can form, and cause video *hum* (furrows in the image).
6. **Lens selection.** Automatic-iris lenses are necessary where changes in lighting will occur. In these cases, a fixed-iris lens will yield poor results.
7. **Scene selection.** The focal length and camera position must be coordinated to serve the camera's intended purpose. An image must show what was intended.

VIDEO SYSTEMS

The principal elements of a CCTV system are these:

- Cameras
- Lenses
- Mountings
- Communication media
- Switching and Synchronizers
- Monitors
- Video Cassette Recorders
- Video Motion Detectors

CAMERAS

It is important that you know that these cameras do not "see" things in the same way as the human eye sees. The human eye adjusts to various focal lengths and light conditions. However, it is possible to use a lens that is larger than the format of the camera. In other words, you can use a one inch lens on a 2/3-inch camera. You cannot, however, use a 2/3-inch lens on a one-inch camera.

The most important factor in choosing lenses is that they must be matched to the area you are trying to monitor. The proper *focal length* (the distance at which the camera is properly focused) and field of view must be coordinated to get the desired results.

There are two formulas which are used to calculate these values. (These formulas are for 2/3-inch formats.)

For calculating the required lens size (measured in millimeters) for monitoring a certain width, the formula is:

$$\frac{8.8 \times \text{distance}}{\text{width}} = \text{focal length}$$

For calculating the required lens size for an area of a certain height, the formula would be:

$$\frac{6.6 \times \text{distance}}{\text{height}} = \text{focal length}$$

It should be noted that lenses are not available in all sizes, and the actual size that will be used will have to be rounded up or down almost always.

We can change these two formulas around to give us the field of view for any size lens. The formula for the field of view would be:

$$\frac{8.8 \times \text{distance}}{\text{focal length}} = \text{field width}$$

The formula for the height of the field of view would be:

$$\frac{6.6 \times \text{distance}}{\text{focal length}} = \text{field width}$$

For 1/2-inch formats, the formula uses 5.9 instead of 8.8 for the width constant, and 4.4 instead of 6.6 for the height constants.

CAMERAS *(cont'd)*

It is recommended that all lenses be equipped with an automatic *iris*, except for those on vidicon cameras (which are used in indoor locations with a more or less constant light level). The iris controls the amount of light entering the camera. This keeps the light level reaching the sensor within acceptable limits. Sensing automatically brings the signal faster than can be noticed. Video cameras cannot do so. They are limited to fixed fields of view, focal lengths, and light sensitivities. Any applications of video cameras must be made with these facts in mind. In fact, in designing a system, it is best to go to the job site with a camera in hand to verify that the cameras actually pick up the desired images and fields of view.

Cameras currently come in four basic varieties. They are as follows:

1. Vidicon Cameras

2. SIT (silicon intensified target) cameras

3. CCDs (charged coupled device) cameras

4. MOS (metal oxide semiconductor) cameras

Vidicon cameras used to be the most commonly used type. They are relatively inexpensive and are well suited to most indoor uses. With normal indoor lighting, vidicon tubes provide an acceptable quality of signal. They are available in 2/3-inch and one-inch formats to accommodate higher and lower levels of indoor light. (The format defines the size of the photosensitive area of the pickup device.) The larger format sensors have greater sensitivity and resolution.

The vidicon is susceptible to damage from very bright lights, and therefore cannot be used in outdoor locations.

SIT cameras are suitable for areas with light levels varying from low levels to bright sunlight. They are often used for outdoor parking lots, underground parking garages, and similar areas. For special applications with extremely low levels of lighting, an intensified version of the SIT camera, called the ISIT, is available.

SIT cameras respond much better to blue light than they do to a yellowish tint. This means that they will work well in areas with mercury or metal halide lighting but will not work nearly as well in areas with high-pressure sodium lighting. (They should not be considered at all for areas with low-pressure sodium lighting, which is purely yellow.)

CCD cameras are an entirely solid-state type of camera. They are now the most popular type, their prices dropping dramatically in the past few years. They have excellent operating characteristics. (They are now so good, that they are sometimes used for TV studio work.) These cameras exhibit no geometric distortion, no lag, and no image retention.

MOS cameras are also solid-state devices, coming in only the 1/2-inch format. MOS cameras do not perform quite as well as the CCD cameras, but they were often used in the past for their economy, being less expensive than CCDs.

Solid-state cameras are quite durable, being able to cope with vibrations and even magnetic fields with very little difficulty.

LENSES

Video lenses come in 1/2-inch, 2/3-inch, and one-inch sizes to match video cameras up or down to the proper level.

Most auto-iris lenses include a *spot filter*, which increases the range of light levels to which the camera will respond. This is usually mandatory for outdoor cameras due to the wide range of light conditions encountered.

SWITCHING AND SYNCHRONIZING

Especially for large installations with so many cameras to monitor, sequential switching of cameras shown on the monitors is essential. If these switchers were not used, the people watching the monitors would be overwhelmed.

There are many sizes of switchers and sequencers having varied capabilities. They range from systems that monitor a few cameras to microprocessor-controlled models that control many cameras through dozens of monitors. The larger the CCTV system, the more necessary switchers and sequencers are.

Switches simply switch the monitor (usually manually) from one camera's signal to another's.

Sequencers automatically switch monitors from one camera to another in a preset sequence. These units come with any number of features. Some of the more useful ones are these:

Character display generators. A character generator displays on screen the identification of the camera whose view is being seen. This can be very useful when monitoring a number of cameras.

Bridging. This feature allows the operator to choose any camera in the sequence for continuous monitoring on a separate monitor. This can be done without interrupting the sequence.

Synchronizing. This feature allows for switching between cameras without a vertical roll, which obviously is undesirable. By synchronizing all of the cameras to the power-line frequency means of electronic circuits which sense the zero crossing point of the AC line voltage, they are synchronized, and the vertical roll is eliminated. This is somewhat more difficult to do with three-phase systems than with single-phase systems. For three-phase systems, a phase adjustable line lock must be used. This feature allows the individual camera to be synchronized to either of the three electrical phases in a three-phase system.

VIDEO CASSETTE RECORDERS

VCRs provide security records that are very important in analyzing an intrusion or other event. There are two types of VCRs that are commonly used for security surveillance systems.

Time-lapse video recorders use microprocessor-based time-lapse techniques to compress long periods of recording time into a short length of tape. The quality is not as good as a standard video tape, but up to 600 hours of recording can be squeezed onto a standard T-120 cassette. Most are equipped with automatic speed switching mechanisms, which increase the recording speed to standard when an alarm is triggered.

Event alarm recorders. These devices have many of the same features as time-lapse recorders, but do not turn on unless an alarm is triggered. After the alarm, they record for a specified period of time and then shut off until the next alarm condition.

Both types of recorders should have built-in character generators which will identify which camera is displaying the area where the alarm condition is sensed.

VIDEO MOTION DETECTORS

One of the more popular accessories to security monitoring systems is a device called a video motion detector. This device, when properly set up, is able to detect irregular motion in the camera's field of view and set off an alarm. Not only will this alert guards to the intrusion, but it may also switch the scene onto a monitor, so it may be constantly monitored. It may also automatically switch on a video cassette recorder and capture all the action on film.

VMDs, however, are not without problems; they are susceptible to false alarms, and can be more trouble than they are worth. Even the movements of the automatic iris can cause trouble. Some VMDs avoid this difficulty by allowing a certain viewing *window* (area of view) to be set. Only irregular activities within that window will set off the alarm. This greatly reduces the possibility of false alarms. For instance, by setting the window on the top of an unused door, the alarm will not be set off by employees walking past the door, while an intruder opening the door will. Analog VMDs typically offer one window per camera, but digital VMDs may offer up to seven windows (at a higher cost).

In addition to using windows to cut out false alarms, many VMDs have various sensitivity settings. By using shorter focal lengths, the sensitivity is reduced. Also, many models have separate sensitivity settings. Another option, called a **retention control**, specifies how long the irregular action must go on before the alarm is triggered. The reduction of false alarms is also helped by the use of high resolution cameras with low noise characteristics.

When using VMDs, there can be no camera motion (Such as panning or zoom features), as such motion would set off the alarm immediately.

OTHER VIDEO COMPONENTS

In addition to the specialized components we have covered in this chapter, CCTV systems use all the standard little black boxes used in the cable TV trade. Most common are:

- **Amplifiers**, which simply increase the strength of the signal.
- **Splitters**, which operate as a "Y" connection, splitting the signal and sending it several different ways.
- **Screen splitters**, which put two or more images on the same screen.
- **Surge suppressors**, which serve to keep power surges, spikes, and noise off the cables connecting the system.

These devices are usually very easy to connect. Simply connect the coaxial cables to a properly specified device as directed by the manufacturer.

DROP PLANT SERVICE CALLS— CABLE TV NETWORKS

Source of calls	Percentage of Calls
Outdoor connectors	8%
Indoor connectors	13%
Indoor cable	21%
Outdoor cable	19%
All other indoor	19%

IMAGE COMPRESSION STANDARDS

Standard	Applications	Transmission rates
MPEG 2	Video	3-15 Mbps
MPEG 1	Video	1.5 to 2 Mbps
JPEG	Still image	64 Kbps to less than 1 M
H.261	Videoconferencing	64 Kbps

ENCODING RATE AND STORAGE REQUIREMENTS (30-SECOND ADVERTISING SPOT)

Transmission bandwidth, Mbps	Storage Mbytes
1.5	6
2.0	8
2.5	10
3.0	12
3.5	14
4.0	16

DIGITAL MODULATION COMPLEXITY

Coding Format	Levels of Coding	Bits/second/Hertz
On-off keying (ASK)	2	1
4 QAM	4	2
8 QAM	8	3
16 QAM	16	4
Frequency shift keying	2	1
Minimum shift keying	2	2
Bipolar phase shift keying	2	1
Quaternary phase shift keying	4	2
8 PSK	8	3
16 PSK	16	4
16 Vestigial Sideband	16	4

PICTURE TUBE DESIGNATIONS

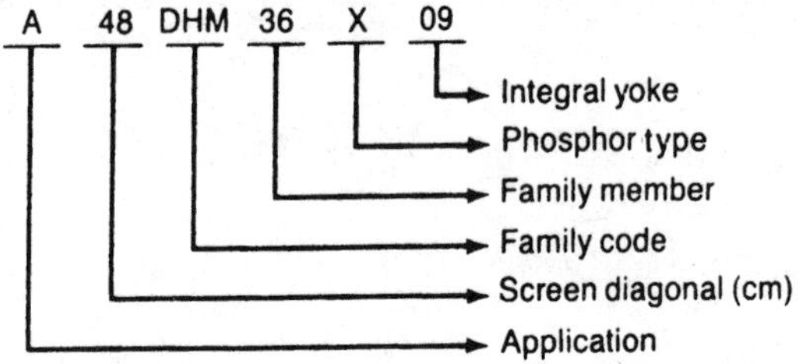

GEOSTATIONARY SATELLITES FROM EAST TO WEST

Position	Name	Inclin.	Drift	Launch	Source
180.0°E	**INTELSAT 701**	0.0°	—	22.10.93	ITSO
178.8°E	**INMARSAT 2-F1**	1.8°	—	30.10.90	IM
177.8°E	**INMARSAT 3-F3**	1.8°	—	18.12.96	IM
177.0°E	**INTELSAT 702**	0.1°	—	17.06.94	ITSO
174.0°E	**INTELSAT 802**	0.0°	—	25.06.97	ITSO

Position	Name	Inclin.	Drift	Launch	Source
169.0°E	**PAS 2**	0.0°	—	08.07.94	US
164.1°E	**OPTUS A2** [AUSSAT 21]	3.7°	—	27.11.85	AUS
162.0°E	**SUPERBIRD B1**	0.0°	—	26.02.92	JPN
160.7°E	**GORIZONT 29**	1.8°	—	18.11.93	CIS

Position	Name	Inclin.	Drift	Launch	Source
160.0°E	**OPTUS B1** [AUSSAT B1]	0.0°	—	13.08.92	AUS
158.0°E	**SUPERBIRD A1**	0.0°	—	01.12.92	JPN
156.0°E	**OPTUS B3**	0.0°	—	27.08.94	AUS
154.0°E	**JCSAT 2**	0.0°	—	01.01.90	JPN
152.0°E	**OPTUS A3** [AUSSAT 3]	1.8°	—	16.09.87	AUS
150.5°E	**PALAPA C1**	0.0°	—	01.02.96	INDO

Position	Name	Inclin.	Drift	Launch	Source
150.0°E	**JCSAT 4**	0.0°	—	17.02.97	JPN
150.0°E	**JCSAT 5**	0.0°	—		
148.1°E	**COSMOS 2155**	3.7°	—	13.09.91	CIS
148.0°E	**MEASAT 2**	0.0°	—	13.11.96	MALA
145.9°E	**AGILA 2**	0.0°	—	19.08.97	RP
145.0°E	**GORIZONT 21**	4.3°	—	03.11.90	CIS
144.0°E	**SUPERBIRD C**	0.0°	—	27.07.97	JPN
140.4°E	**GMS 5**	0.4°	—	18.03.95	JPN
140.0°E	**GORIZONT 22**	4.3°	—	23.11.90	CIS

GEOSTATIONARY SATELLITES FROM EAST TO WEST *(cont'd)*

Position	Name	Inclin.	Drift	Launch	Source
137.9°E	**APSTAR 1**	0.0°	—	21.07.94	PRC
135.9°E	**NSTAR 2**	0.0°	—	05.02.96	JPN
133.9°E	**APSTAR 1A**	0.0°	—	03.07.96	PRC
132.0°E	**NSTAR 1**	0.0°	—	29.08.95	JPN

Position	Name	Inclin.	Drift	Launch	Source
128.0°E	**JCSAT 3**	0.0°	—	29.08.95	JPN
127.0°E	**RADUGA 27**	4.3°	—	28.02.91	CIS
124.9°E	**DFH-32**	0.1°	—	11.05.97	PRC
122.2°E	**GORIZONT 30**	1.5°	—	20.05.94	CIS

Position	Name	Inclin.	Drift	Launch	Source
119.9°E	**THAICOM 1**	0.1°	—	18.12.93	THAI
119.9°E	**HIMAWARI 4** [GMS 4]	3.0°	—	05.09.89	JPN
117.9°E	**PALAPA B4**	0.0°	—	14.05.92	INDO
116.0°E	**KOREASAT 2**	0.0°	—	14.01.96	KOR
115.9°E	**KOREASAT 1**	0.1°	—	05.08.95	KOR
115.6°E	**SPACENET 2**	0.7°	—	10.11.84	US
115.5°E	**CHINASAT 5** [SPACENET 1]	1.6°	—	23.05.84	PRC
113.0°E	**PALAPA C2**	0.0°	—	16.05.96	INDO
111.0°E	**JCSAT 1**	0.5°	—	06.03.89	JPN
110.9°E	**STTW-3**	2.2°	—	22.12.88	PRC
110.0°E	**BSAT 1A**	0.0°	—	16.04.97	JPN

Position	Name	Inclin.	Drift	Launch	Source
109.9°E	**BS-3B** [YURI 3B]	0.0°	—	25.08.91	JPN
109.7°E	**BS-3N**	0.1°	—	08.07.94	JPN
108.0°E	**BS-3A** [YURI 3A]	0.0°	—	28.08.90	JPN
107.6°E	**INDOSTAR 1**	0.1°	—		
107.5°E	**PALAPA B2R**	0.0°	0.1°W	13.04.90	INDO
105.4°E	**ASIASAT 1**	0.0°	—	07.04.90	AC
104.8°E	**FY-2**	0.7°	—	10.06.97	PRC
102.7°E	**GORIZONT 25**	3.1°	—	02.04.92	CIS
100.5°E	**ASIASAT 2**	0.0°	—	28.11.95	AC

GEOSTATIONARY SATELLITES FROM EAST TO WEST *(cont'd)*

Position	Name	Inclin.	Drift	Launch	Source
98.8°E	**EKRAN 20**	2.6°	—	30.10.92	CIS
98.4°E	**STTW-4**	2.0°	—	04.02.90	PRC
96.6°E	**GORIZONT 27**	2.6°	—	27.11.92	CIS
93.5°E	**INSAT 2B**	0.1°	—	22.07.93	IND
93.5°E	**INSAT 2C**	0.1°	—	06.12.95	IND
91.5°E	**MEASAT 1**	0.0°	—	12.01.96	MALA
90.2°E	**GORIZONT 28**	1.8°	—	28.10.93	CIS

Position	Name	Inclin.	Drift	Launch	Source
85.4°E	**RADUGA 30**	1.9°	—	30.09.93	CIS
84.9°E	**TDRS 3**	3.7°	—	29.09.88	US
83.0°E	**INSAT 1D**	0.1°	—	12.06.90	IND
80.0°E	**COSMOS 2319**	0.4°	—	30.08.95	CIS

Position	Name	Inclin.	Drift	Launch	Source
80.0°E	**EXPRESS 2**	0.1°	—	26.09.96	CIS
79.8°E	**GORIZONT 24**	3.5°	—	23.10.91	CIS
78.5°E	**THAICOM 3**	0.0°	—	16.04.97	THAI
78.5°E	**THAICOM 2**	0.1°	—	08.10.94	THAI
77.6°E	**LEASAT 5**	4.0°	—	09.01.90	US
76.6°E	**LUCH 1**	1.2°	—	11.10.95	CIS
76.5°E	**STTW-2**	3.7°	0.1°W	07.03.88	PRC
76.4°E	**APSTAR 2R**	0.0°	—	16.10.97	PRC
76.3°E	**ELEKTRO** [GOMS]	1.3°	—	31.10.94	CIS
75.7°E	**RADUGA 26**	4.2°	0.2°W	20.12.90	CIS
74.2°E	**COSMOS 2133**	3.0°	0.1°W	14.02.91	CIS
74.0°E	**RADUGA 1-2**	4.1°	0.2°W	27.12.90	CIS
73.9°E	**INSAT 2A**	0.6°	—	09.07.92	IND
71.3°E	**COSMOS 2085**	4.5°	—	18.07.90	CIS
71.1°E	**UFO 2** [USA 95]	3.5°	—	03.09.93	US
70.5°E	**RADUGA 32**	0.9°	—	28.12.94	CIS

GEOSTATIONARY SATELLITES FROM EAST TO WEST *(cont'd)*

Position	Name	Inclin.	Drift	Launch	Source
68.5°E	**PAS 4**	0.0°	—	03.08.95	US
66.1°E	**INTELSAT 704**	0.0°	—	10.01.95	ITSO
65.3°E	**RADUGA 31**	1.5°	0.2°E	18.02.94	CIS
64.9°E	**INMARSAT 2-F3**	0.6°	—	16.12.91	IM
64.1°E	**INTELSAT 804**	0.0°	—		
63.8°E	**INMARSAT 3-F1**	1.3°	—	03.04.96	IM
62.0°E	**INTELSAT 602**	0.0°	—	27.10.89	ITSO

Position	Name	Inclin.	Drift	Launch	Source
60.0°E	**INTELSAT 604**	0.0°	—	23.06.90	ITSO
56.9°E	**INTELSAT 703**	0.0°	—	06.10.94	ITSO
55.0°E	**KUPON**	0.0°	—		
54.9°E	**ARABSAT 1C**	0.0°	—	26.02.92	AB
53.0°E	**SKYNET 4B**	4.5°	—	11.12.88	UK
53.0°E	**GORIZONT 32**	0.2°	—	25.05.96	CIS

Position	Name	Inclin.	Drift	Launch	Source
48.8°E	**RADUGA 1-3**	1.6°	—	05.02.94	CIS
42.0°E	**TURKSAT 1C**	0.0°	—	09.07.96	TURK

Position	Name	Inclin.	Drift	Launch	Source
39.7°E	**GORIZONT 31**	0.2°	—	25.01.96	CIS
36.1°E	**GALS 2**	0.2°	—	17.11.95	CIS
36.0°E	**GALS 1**	0.3°	—	20.01.94	CIS
35.6°E	**TDF 2**	0.0°	—	24.07.90	FR
35.0°E	**RADUGA 28**	3.3°	—	19.12.91	CIS
33.1°E	**INTELSAT 510**	4.3°	—	22.03.85	ITSO
31.3°E	**TURKSAT 1B**	0.0°	—	10.08.94	TURK
30.5°E	**ARABSAT 2B**	0.0°	—	13.11.96	AB

GEOSTATIONARY SATELLITES FROM EAST TO WEST *(cont'd)*

Position	Name	Inclin.	Drift	Launch	Source
28.5°E	**DFS 2**	0.0°	—	24.07.90	FRG
27.3°E	**INMARSAT 3-F5**	2.8°	—		
25.9°E	**ARABSAT 2A**	0.0°	—	09.07.96	AB
25.7°E	**GORIZONT 20**	4.6°	—	20.06.90	CIS
25.5°E	**EUTELSAT 1-F4** [ECS 4]	3.7°	—	16.09.87	EUTE
23.4°E	**DFS 3**	0.0°	—	12.10.92	FRG
21.5°E	**EUTELSAT 1-F5** [ECS 5]	2.9°	—	21.07.88	EUTE

Position	Name	Inclin.	Drift	Launch	Source
19.2°E	**ASTRA 1F**	0.0°	—	08.04.96	LUXE
19.2°E	**ASTRA 1G**	0.0°	—		
19.2°E	**ASTRA 1E**	0.1°	—	19.10.95	LUXE
19.2°E	**ASTRA 1C**	0.0°	—	12.05.93	LUXE
19.1°E	**ASTRA 1B**	0.0°	—	02.03.91	LUXE
19.1°E	**ASTRA 1D**	0.0°	0.1°E	01.11.94	LUXE
19.1°E	**ASTRA 1A**	0.0°	—	11.12.88	LUXE
16.0°E	**EUTELSAT 2-F3**	0.0°	—	07.12.91	EUTE
13.1°E	**ITALSAT 1**	0.8°	—	15.01.91	IT
13.1°E	**ITALSAT 2**	0.8°	—	08.08.96	IT
12.9°E	**HOTBIRD 2**	0.1°	—	21.11.96	EUTE
12.9°E	**EUTELSAT 2-F1**	0.1°	—	30.08.90	EUTE
12.9°E	**EUTELSAT II-F6X** [HOTBIRD 1]	0.1°	—	28.03.95	ELITE
12.9°E	**HOTBIRD 3**	0.1°	—	02.09.97	EUTE
12.3°E	**RADUGA 29**	2.3°	0.1°W	25.03.93	CIS
12.0°E	**COSMOS 2224**	1.7°	—	17.12.92	CIS
10.0°E	**EUTELSAT 2-F2**	0.0°	—	15.01.91	EUTE

GEOSTATIONARY SATELLITES FROM EAST TO WEST *(cont'd)*

Position	Name	Inclin.	Drift	Launch	Source
6.9°E	**EUTELSAT 2-F4**	0.0°	—	09.07.92	EUTE
5.2°E	**SIRIUS** [MARCOPOLO 1]	0.0°	—	27.08.89	SWED
4.9°E	**SKYNET 4D**	4.1°	—		
4.8°E	**SIRIUS 2**	0.0°	—		
2.9°E	**TELECOM 2C**	0.0°	—	06.12.95	FR

Position	Name	Inclin.	Drift	Launch	Source
0.0°W	**METEOSAT 6**	0.2°	—	20.11.93	ESA
0.5°W	**TVSAT 2**	0.1°	—	08.08.89	FRG
0.6°W	**THOR 1** [MARCOPOLO 2]	0.0°	—	18.08.90	NOR
0.8°W	**THOR 2A**	0.0°	—	20.05.97	NOR
1.0°W	**SKYNET 4C**	1.5°	—	30.08.90	UK
1.1°W	**INTELSAT 707**	0.0°	—	14.03.96	ITSO
4.0°W	**AMOS**	0.1°	—	16.05.96	ISRA
5.1°W	**TELECOM 2B**	0.0°	—	15.04.92	FR
5.1°W	**TELECOM 2D**	0.1°	—	08.08.96	FR
8.0°W	**TELECOM 2A**	0.0°	—	16.12.91	FR

Position	Name	Inclin.	Drift	Launch	Source
10.1°W	**METEOSAT 7**	1.5°	—	02.09.97	ESA
10.9°W	**GORIZONT 26**	2.8°	—	14.07.92	CIS
13.7°W	**COSMOS 2291**	1.1°	—	21.09.94	CIS
14.0°W	**EKSPRESS 1**	0.0°	—	13.10.94	CIS
15.7°W	**INMARSAT 3-F2**	1.7°	—	06,09.96	IM
15.8°W	**UFO 3** [USA 104]	3.5°	—	24.06.94	US
16.4°W	**LUCH**	0.3°	—	16.12.94	CIS
16.9°W	**INMARSAT 2-F4**	2.6°	—	15.04.92	IM
17.8°W	**NATO 4A**	2.1°	—	08.01.91	NATO
18.1°W	**INTELSAT 705**	0.0°	—	22.03.95	ITSO

GEOSTATIONARY SATELLITES FROM EAST TO WEST *(cont'd)*

Position	Name	Inclin.	Drift	Launch	Source
21.5°W	**INTELSAT K**	0.0°	—	10.06.92	ITSO
21.6°W	**INTELSAT 803**	0.0°	—	23.09.97	ITSO
22.0°W	**FLTSATCOM 8** [USA 46]	2.2°	—	25.09.89	US
23.7°W	**UFO 7** [USA 127]	4.4°	—	25.07.96	US
24.6°W	**INTELSAT 603**	0.0°	—	14.03.90	ITSO
27.0°W	**COSMOS 2345**	0.9°	—	14.08.97	CIS
27.6°W	**INTELSAT 605**	0.0°	—	14.08.91	ITSO
30.0°W	**HISPASAT 1B**	0.0°	—	22.07.93	SPN

Position	Name	Inclin.	Drift	Launch	Source
30.0°W	**HISPASAT 1A**	0.0°	—	10.09.92	SPN
34.0°W	**SKYNET 4A**	3.2°	—	01.01.90	UK
34.5°W	**INTELSAT 601**	0.0°	—	29.10.91	ITSO
35.1°W	**INTELSAT 515**	0.5°	0.3°W	27.01.89	ITSO
37.5°W	**ORION 1**	0.0°	—	29.11.94	US

Position	Name	Inclin.	Drift	Launch	Source
40.8°W	**COSMOS 2054**	5.0°	0.2°W	27.12.89	CIS
41.0°W	**TDRS 4**	1.1°	—	13.03.89	US
43.0°W	**PAS 3R**	0.0°	—	12.01.96	US
43.3°W	**PAS 6**	0.1°	—	07.08.97	US
45.0°W	**PAS 1**	0.1°	—	15.06.88	US
47.0°W	**TDRS F6**	0.0°	—	13.01.93	US

Position	Name	Inclin.	Drift	Launch	Source
50.0°W	**INTELSAT 709**	0.0°	—	15.06.96	ITSO
53.1°W	**INTELSAT 706**	0.0°	—	17.05.95	ITSO
54.1°W	**INMARSAT 3-F4**	0.0°	—	03.06.97	IM
55.2°W	**INMARSAT 2-F2**	2.1°	—	08.03.91	IM
55.5°W	**INTELSAT 12**	3.1°	—	28.09.85	ITSO
58.1°W	**PAS 5**	0.0°	—	28.08.97	US

GEOSTATIONARY SATELLITES FROM EAST TO WEST *(cont'd)*

Position	Name	Inclin.	Drift	Launch	Source
61.5°W	**ECHOSTAR 3**	0.1°	—	05.10.97	US
65.0°W	**BRAZILSAT B2**	0.0°	—	28.03.95	BRAZ
65.1°W	**BRAZILSAT B3**	0.0°	—		

Position	Name	Inclin.	Drift	Launch	Source
70.1°W	**BRAZILSAT B1**	0.0°	—	10.08.94	BRAZ
71.9°W	**NAHUEL 1A**	0.0°	—	30.01.97	ARGN
74.1°W	**GALAXY 6**	0.0°	—	12.10.90	US
74.1°W	**SBS 6**	0.0°	—	12.10.90	US
75.7°W	**GOES 8**	0.3°	—	13.04.94	US
77.0°W	**SBS 4**	3.9°	—	30.08.84	US
79.0°W	**BRAZILSAT 1**	2.6°	—	08.02.85	BRAZ

Position	Name	Inclin.	Drift	Launch	Source
81.0°W	**SATCOM K2**	0.7°	—	27.11.95	US
83.0°W	**SPACENET 3R**	0.0°	—	11.03.88	US
84.9°W	**GE 2**	0.0°	—	30.01.97	US
87.1°W	**GE 3**	0.1°	—	04.09.97	US
89.1°W	**TELSTAR 402R**	0.1°	—	24.09.95	US

Position	Name	Inclin.	Drift	Launch	Source
91.0°W	**GALAXY 7**	0.0°	—	28.10.92	US
91.9°W	**ASC 1**	2.7°	0.1°E	27.08.85	US
92.0°W	**BRAZILSAT 2**	1.2°	—	28.03.86	BRAZ
95.1°W	**GALAXY 3R**	0.0°	—	15.12.95	US
97.0°W	**TELSTAR 5**	0.0°	—	24.05.97	US
97.8°W	**GOES 7**	4.3°	—	26.02.87	US
99.0°W	**GALAXY 4**	0.0°	—	25.06.93	US
99.6°W	**FLTSATCOM 7** [USA 20]	4.0°	—	05.12.86	US

GEOSTATIONARY SATELLITES FROM EAST TO WEST *(cont'd)*

Position	Name	Inclin.	Drift	Launch	Source
100.0°W	**ACTS**	0.1°	—	12.09.93	US
100.8°W	**DBS 2**	0.0°	—	03.08.94	US
100.9°W	**DBS 3**	0.0°	—	10.06.95	US
101.0°W	**AMSC 1**	0.0°	—	07.04.95	US
101.1°W	**SPACENET 4** (ASC 2]	0.0°	—	13.04.91	US
101.2°W	**DBS 1**	0.0°	—	18.12.93	US
103.0°W	**GE 1**	0.0°	—	08.09.96	US
105.0°W	**GSTAR 4**	0.0°	—	20.11.90	US
105.2°W	**GSTAR 1**	1.4°	—	08.05.85	US
106.0°W	**UFO 6** [USA 114]	4.2°	—	22.10.95	US
106.1°W	**ISO**	2.1°	0.2°W	17.11.95	ESA
106.5°W	**MSAT M1**	0.0°	—	20.04.96	CA
107.3°W	**GOES 10**	0.0°	—	25.04.97	US
107.4°W	**ANIK E2**	0.0°	—	04.04.91	CA
109.2°W	**SOLIDARIDAD 1**	0.0°	—	20.11.93	MEX

Position	Name	Inclin.	Drift	Launch	Source
111.2°W	**ANIK E1**	0.1°	—	26.09.91	CA
112.6°W	**TELSTAR 401**	0.9°	—	16.12.93	US
112.9°W	**GALAXY 2**	3.0°	0.5°W	22.09.83	US
113.0°W	**SOLIDARIDAD 2**	0.0°	—	08.10.94	MEX
116.8°W	**MORELOS 2**	0.1°	—	27.11.85	MEX
118.7°W	**NAHUEL 1** [ANIK C1]	0.8°	—	12.04.85	ARGN
118.8°w	**TEMPO 2**	0.1°	—	08.03.97	US
118.9°W	**ECHOSTAR 1**	0.0°	—	28.12.95	US
119.1°W	**ECHOSTAR 2**	0.0°	—	11.09.96	US

GEOSTATIONARY SATELLITES FROM EAST TO WEST *(cont'd)*

Position	Name	Inclin.	Drift	Launch	Source
120.0°W	**TELSTAR 303**	2.1°	—	17.06.85	US
123.0°W	**SBS 5**	0.0°	—	08.09.88	US
123.1°W	**GALAXY 9**	0.0°	—	24.05.96	US
125.1°W	**GALAXY 5**	0.0°	—	14.03.92	US

Position	Name	Inclin.	Drift	Launch	Source
130.9°W	**SATCOM C3**	0.0°	—	10.09.92	US
133.0°W	**GALAXY 1R**	0.0°	—	19.02.94	US
135.1°W	**SATCOM C4**	0.0°	—	31.08.92	US
135.5°W	**GOES 9**	0.1°	—	23.05.95	US
137.0°W	**SATCOM C1**	0.0°	—	20.11.90	US
139.0°W	**AURORA 2**	0.0°	—	29.05.91	US

Position	Name	Inclin.	Drift	Launch	Source
144.7°W	**COSMOS 2209**	2.9°	0.3°W	10.09.92	CIS

Position	Name	Inclin.	Drift	Launch	Source
162.6°W	**COSMOS 2172**	3.4°	0.2°W	22.11.91	CIS
168.6°W	**COSMOS 2282**	0.7°	0.1°E	06.07.94	CIS

Position	Name	Inclin.	Drift	Launch	Source
170.3°W	**RADUGA 25**	4.9°	—	15.02.90	CIS
171.3°W	**TDRS F7**	1.9°	—	13.07.95	US
174.4°W	**TDRS 5**	0.2°	—	02.08.91	US
177.1°W	**INTELSAT 513**	2.0°	—	17.05.88	ITSO
177.9°W	**UFO 4** [USA 108]	4.0°	0.1°W	29.01.95	US

DSB-TV IN EUROPE				
YEAR	SATELLITE SYSTEM	FORMAT	CHANNEL POWER OR EIRP (dBW)	COMMENTS
Current	Various	TV	10 W, typical	FSS bands used
1989	Olympus	TV	180 W	ESA, experimental, Ku-band
1989	TV SAT	TV	230-260 W	France & Germany, operational, Ku-band
1992	HISP ASAT	TV	110w	Spain
1994	PRE-EUROPESAT	TV		EUTELSAT
1995, 96 & 97	EUROPESAT (2 Ops. plus spare)	TV Digital HDTV	75 W 55-60 dBW	EUTELSAT, 80–90 cm receive antenna
1995	EUREKA 95 (Terrestrial & Satellite	HD-MAC		Pan-European program
	Flash TV (Terrestrial and Satellite)	High Quality Std. for theaters, etc.		EC, 2–4m receive antenna

DSB-TV IN JAPAN				
YEAR	SATELLITE SYSTEM	FORMAT	CHANNEL POWER OR EIRP (dBW)	COMMENTS
1978	BSE	TV	100 W	Experimental
1984	BS-2a	TV	100 W	Operational system
1986	BS-2b	TV	100 W	2 Channel, 24 hour service in 1989
1990 & 1991	BS-3a & 3b	TV, MUSE-TV, Hi-Vision (HDTV)	120 W 59 dBW	Operational systems
1996	COMETS	HDTV	200 W ~65 dBW	22 GHZ experimental

CHAPTER 8
COMPUTER AND INTERNET

SCSI PORT SINGLE-ENDED 50-PIN PIN-OUT ASSIGNMENTS

Pin Number	Signal Name
2	–DB0
4	–DB1
6	–DB2
8	–DB3
10	–DB4
12	–DB5
14	–DB6
16	–DB7
18	–DBP
20	GROUND
22	GROUND
24	GROUND
26	TERMPWR
28	GROUND
30	GROUND
32	–ATN
34	GROUND
36	–BSY
38	–ACK
40	–RST
42	–MSG
44	–SEL
46	–C/D
48	–REQ
50	–I/O

SCSI PORT DIFFERENTIAL 50-PIN PIN-OUT ASSIGNMENTS

Pin Number	Signal Name	Pin Number	Signal Name
1	SHIELD/GROUND	26	TERMPWR
2	GROUND	27	GROUND
3	+DB0	28	GROUND
4	−DB0	29	+ATN
5	+DB1	30	−ATN
6	−DB1	31	GROUND
7	+DB2	32	GROUND
8	−DB2	33	+BSY
9	+DB3	34	−BSY
10	−DB3	35	+ACK
11	+DB4	36	−ACK
12	−DB4	37	+RST
13	+DB5	38	−RST
14	−DB5	39	+MSG
15	+DB6	40	−MSG
16	−DB6	41	+SEL
17	+DB7	42	−SEL
18	−DB7	43	+C/D
19	+DBP	44	−C/D
20	−DBP	45	+REQ
21	DIFFSENS	46	−REQ
22	GROUND	47	+I/O
23	GROUND	48	−I/O
24	GROUND	49	GROUND
25	TERMPWR	50	GROUND

SINGLE-ENDED DB-25 PIN-OUT ASSIGNMENTS

Pin Number	Signal Name
1	–REQ
2	–MSG
3	–I/O
4	–RST
5	–ACK
6	–BSY
7	GROUND
8	–DB0
9	GROUND
10	–DB3
11	–DB5
12	–DB6
13	–DB7
14	GROUND
15	–C/D
16	GROUND
17	–ATN
18	GROUND
19	–SEL
20	–DBP
21	–DB1
22	–DB2
23	–DB4
24	GROUND
25	TERMPWR

SIGNALS ON A CENTRONICS-TYPE CONNECTOR

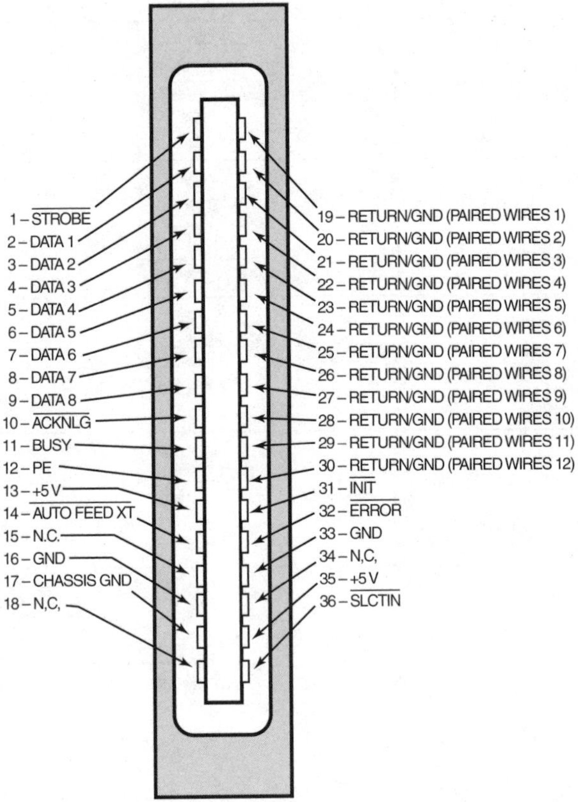

1 – STROBE
2 – DATA 1
3 – DATA 2
4 – DATA 3
5 – DATA 4
6 – DATA 5
7 – DATA 6
8 – DATA 7
9 – DATA 8
10 – ACKNLG
11 – BUSY
12 – PE
13 – +5 V
14 – AUTO FEED XT
15 – N.C.
16 – GND
17 – CHASSIS GND
18 – N,C,
19 – RETURN/GND (PAIRED WIRES 1)
20 – RETURN/GND (PAIRED WIRES 2)
21 – RETURN/GND (PAIRED WIRES 3)
22 – RETURN/GND (PAIRED WIRES 4)
23 – RETURN/GND (PAIRED WIRES 5)
24 – RETURN/GND (PAIRED WIRES 6)
25 – RETURN/GND (PAIRED WIRES 7)
26 – RETURN/GND (PAIRED WIRES 8)
27 – RETURN/GND (PAIRED WIRES 9)
28 – RETURN/GND (PAIRED WIRES 10)
29 – RETURN/GND (PAIRED WIRES 11)
30 – RETURN/GND (PAIRED WIRES 12)
31 – INIT
32 – ERROR
33 – GND
34 – N,C,
35 – +5 V
36 – SLCTIN

PIN ASSIGNMENTS FOR AN RS-232C CONNECTION

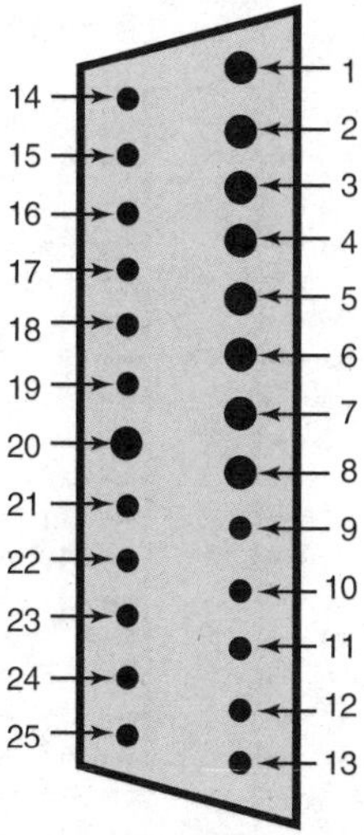

1. Ground	14. Secondary Transmitted Data
2. Transmitted Data	15. Transmit Clock
3. Received Data	16. Secondary Received Data
4. Request To Send	17. Receiver Clock
5. Clear To Send	18. Unassigned
6. Data Set Ready	19. Secondary Request To Send
7. Logic Ground	20. Data Terminal Ready
8. Carrier Detect	21. Signal Quality Detect
9. Reserved	22. Ring Detect
10. Reserved	23. Data Rate Select
11. Unassigned	24. Transmit Clock
12. Secondary Carrier	25. Unassigned
13. Secondary Clear	

PIN-OUT ARRANGEMENT CARD CONNECTOR SLOT

GND–B1 ▬ ▬ A1–NMI
RESET–B2 ▬ ▬ A2–D7
+5 V–B3 ▬ ▬ A3–D6
IA2–B4 ▬ ▬ A4–D5
–5 V–B5 ▬ ▬ A5–D4
$\overline{\text{FDCMRQ}}$–B6 ▬ ▬ A6–D3
–12 V–B7 ▬ ▬ A7–D2
AUDIO IN–B8 ▬ ▬ A8–D1
+12 V–B9 ▬ ▬ A9–D0
$\overline{\text{GND}}$–B10 ▬ ▬ A10–RDYIN
$\overline{\text{MEMW}}$–B11 ▬ ▬ A11–AEN
$\overline{\text{MEMR}}$–B12 ▬ ▬ A12–A19
$\overline{\text{IOW}}$–B13 ▬ ▬ A13–A18
$\overline{\text{IOR}}$–B14 ▬ ▬ A14–A17
$\overline{\text{DACK3}}$–B15 ▬ ▬ A15–A16
$\overline{\text{DRQ3}}$–B16 ▬ ▬ A16–A15
$\overline{\text{DACK1}}$–B17 ▬ ▬ A17–A14
$\overline{\text{DRQ1}}$–B18 ▬ ▬ A18–A13
$\overline{\text{REFRESH}}$–B19 ▬ ▬ A19–A12
CLOCK–B20 ▬ ▬ A20–A11
IA7–B21 ▬ ▬ A21–A10
IA6–B22 ▬ ▬ A22–A9
IA5–B23 ▬ ▬ A23–A8
IA4–B24 ▬ ▬ A24–A7
IA3–B25 ▬ ▬ A25–A6
$\overline{\text{FDCDACK}}$–B26 ▬ ▬ A26–A5
DMATC–B27 ▬ ▬ A27–A4
ALE–B28 ▬ ▬ A28–A3
+5 V–B29 ▬ ▬ A29–A2
OSC–B30 ▬ ▬ A30–A1
GND–B31 ▬ ▬ A31–A0

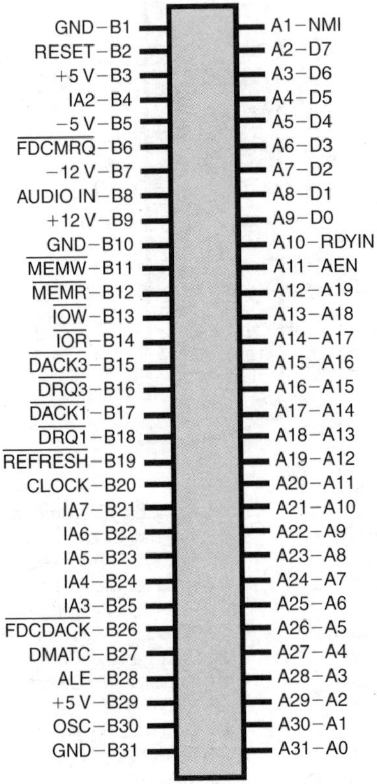

ASCII CODE CHART

ASCII Code Chart

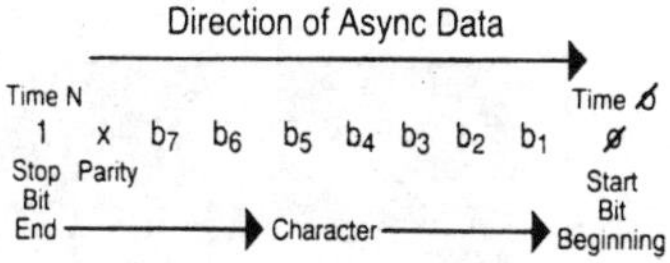

b_4	b_3	b_2	b_1	MSB / LSB								
				b_7	0	0	0	0	1	1	1	1
				b_6	0	0	1	1	0	0	1	1
				b_5	0	1	0	1	0	1	0	1
				row \ colm	**0**	**1**	**2**	**3**	**4**	**5**	**6**	**7**
0	0	0	0	**0**	NUL	DLE	SP	0	@	P	`	p
0	0	0	1	**1**	SOH	DC_1	!	1	A	Q	a	q
0	0	1	0	**2**	STX	DC_2	"	2	B	R	b	r
0	0	1	1	**3**	ETX	DC_3	#	3	C	S	c	s
0	1	0	0	**4**	EOT	DC_4	$	4	D	T	d	t
0	1	0	1	**5**	ENQ	NAK	%	5	E	U	e	u
0	1	1	0	**6**	ACK	SYN	&	6	F	V	f	v
0	1	1	1	**7**	BEL	ETB	'	7	G	W	g	w
1	0	0	0	**8**	BS	CAN	(	8	H	X	h	x
1	0	0	1	**9**	HT	EM	)	9	I	Y	i	y
1	0	1	0	**A**	LF	SUB	*	:	J	Z	j	z
1	0	1	1	**B**	VT	ESC	+	;	K	[	k	{
1	1	0	0	**C**	FF	FS	,	<	L	\	l	¦
1	1	0	1	**D**	CR	GS	-	=	M	]	m	}
1	1	1	0	**E**	SO	RS	.	>	N	^	n	~
1	1	1	1	**F**	SI	US	/	?	O	_	o	DEL

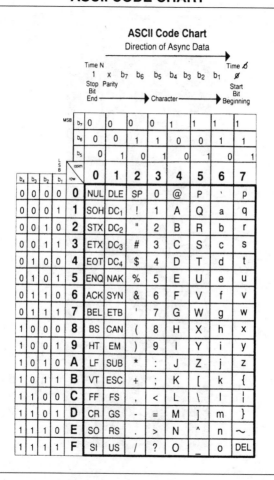

V.35 INTERFACE

The V.35 interface is typically found on DTE and DCE equipment interfacing to high speed digital carrier services.

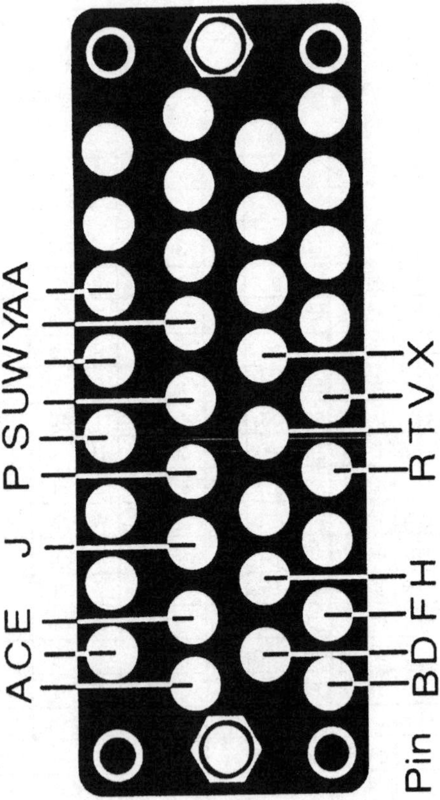

V.35 INTERFACE *(cont'd)*

Pin	Signal	Pin	Signal
A	Chassis Ground	B	Signal Ground
C	Request to Send	D	Clear to Send
E	Data Set Ready	F	Rcv. Line Signal Defect
H	Data Terminal Ready	J	Ring Indicator
P	Transmitted Data (Sig. A)	R	Received Data (Sig. A)
S	Transmitted Data (Sig. B)	T	Received Data (Sig. B)
U	Terminal Timing	V	Receive Timing A
W	Terminal Timing	X	Receive Timing
Y	Terminal Timing	AA	Transmit Timing

PARALLEL INTERFACE (CENTRONICS TYPE)

The Centronics® Parallel Interface is a 36 pin, byte-serial interface which has become a widely accepted standard used in computer printer communications.

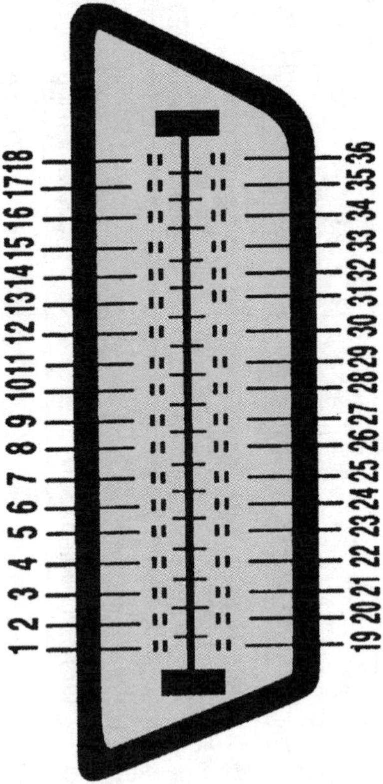

PARALLEL INTERFACE (CENTRONICS TYPE) *(cont'd)*

Pin	Signal	Pin	Signal
1	Data Strobe	19	(R) Data Strobe
2	Data Bit 1	20	(R) Data Bit 1
3	Data Bit 2	21	(R) Data Bit 2
4	Data Bit 3	22	(R) Data Bit 3
5	Data Bit 4	23	(R) Data Bit 4
6	Data Bit 5	24	(R) Data Bit 5
7	Data Bit 6	25	(R) Data Bit 6
8	Data Bit 7	26	(R) Data Bit 7
9	Data Bit 8	27	(R) Data Bit 8
10	Acknowledge	28	(R) Acknowledge
11	Busy	29	(R) Busy
12	Paper End	30	(R) Paper End
13	Select	31	Input Prime
14	Supply Ground	32	Fault
15	OSCXT	33	Undefined
16	Logic Ground	34	Undefined
17	Chassis Ground	35	Undefined
18	+5V	36	Undefined

X.21 INTERFACE

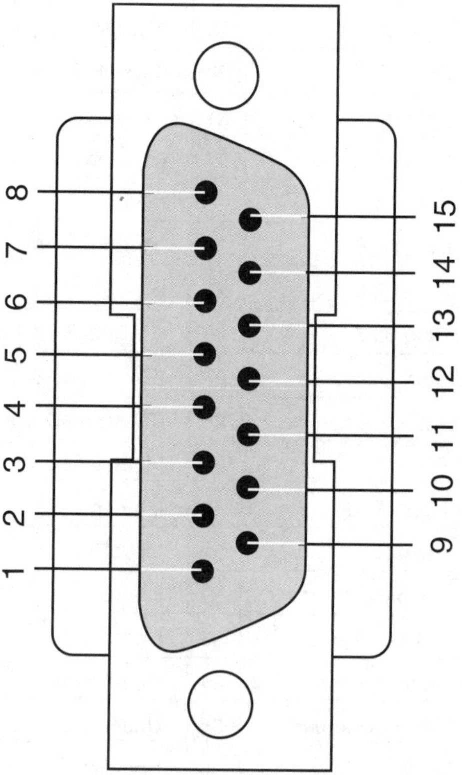

X.21 INTERFACE *(cont'd)*

Pin	Signal	Pin	Signal
1	Shield	9	Transmit (B)
2	Transmit (A)	10	Control (B)
3	Control (A)	11	Receive (B)
4	Receive (A)	12	Indication (B)
5	Indication (A)	13	Signal Timing (B)
6	Signal Timing (A)	14	Unassigned
7	Unassigned	15	Unassigned
8	Ground (GND)		

EIA-530 INTERFACE

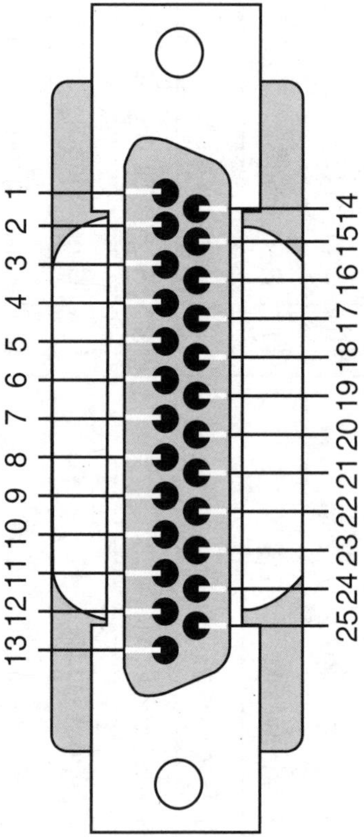

EIA-530 INTERFACE *(cont'd)*

Pin	Signal	Pin	Signal
1	Shield	14	Transmitted Data Return
2	Transmitted Data	15	Transmit Signal Element Timing
3	Received Data	16	Received Data Return
4	Request to Send	17	Rec. Sig. Element Timer Return
5	Clear to Send	18	Local Loopback
6	DCE Ready	19	Request to Send
7	Signal Ground	20	DTE Ready
8	Received Line Signal Detector	21	Remote Loopback
9	Receiver Signal Element Timer	22	DCE Ready
10	Received Line Signal Detector	23	DTE Ready
11	Transmit Signal Element Timing	24	Trans. Sig. Element Timing Return
12	Transmit Signal Element Timing	25	Test Mode
13	Clear to Send		

RS-232 INTERFACE

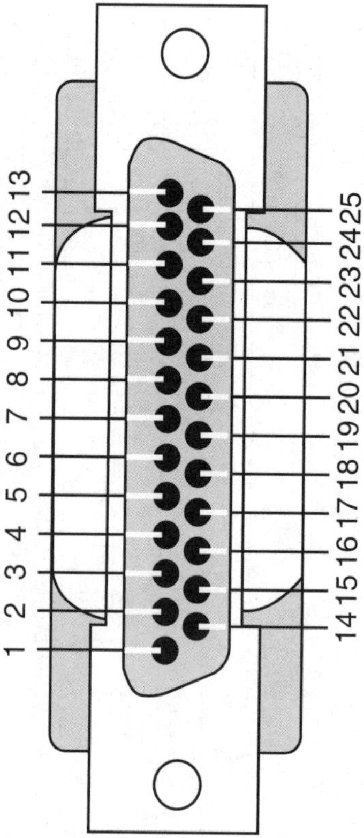

RS-232 INTERFACE *(cont'd)*

Pin	Description	EIA CKT	From DCE	To DCE
1	Frame Ground	AA		
2	Transmitted Data	BA		D (Data)
3	Received Data	BB	D	
4	Request to Send	CA		C (Control)
5	Clear to Send	CB	C	
6	Data Set Ready	CC	C	
7	Signal Gnd/Common Return	AB		
8	Rcvd. Line Signal Detector	CF	C	
11	Undefined			
12	Secondary Rcvd. Line Sig. Detector	SCF	C	
13	Secondary Clear to Send	SCB	C	
14	Secondary Transmitted Data	SBA		D
15	Transmitter Sig. Element Timing	DB	T (Timing)	
16	Secondary Received Data	SBB	D	
17	Receiver Sig. Element Timing	DD	T	
18	Undefined			
19	Secondary Request to Send	SCA		C
20	Data Terminal Ready	CD		C
21	Sig. Quality Detector	CG		C
22	Ring Indicator	CE	C	
23	Data Sig. Rate Selector (DCE)	CI	C	
23	Data Sig. Rate Selector (DTE)	CH		C
24	Transmitter Sig. Element Timing	DA		T
25	Undefined			

RS-449 INTERFACE

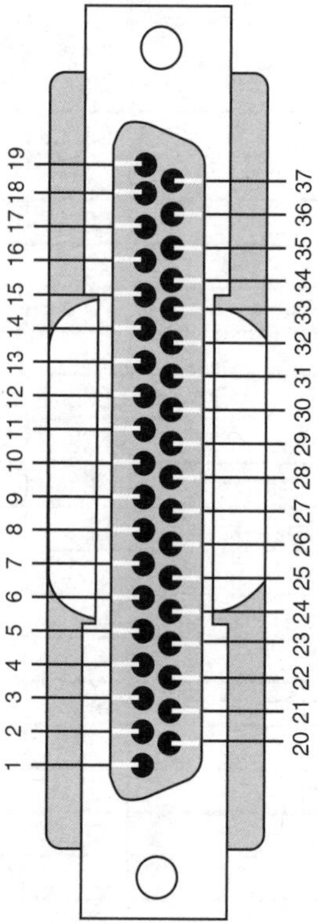

RS-449 INTERFACE *(cont'd)*

Pin A	Pin B	EIA CKT	Description	From DCE	To DCE
1			Shield		
2		SI	Signaling Rate Indicator	*C	
4	22	SD	Send Data		*D
5	23	ST	Send Timing	*T	
6	24	RD	Receive Data	*D	
7	25	RS	Request to Send		*C
8	26	RT	Receive Timing	*T	
9	27	CS	Clear to Send	*C	
10		LL	Local Loopback		*C
11	29	DM	Data Mode	*C	
12	30	TR	Terminal Ready		*C
13	31	RR	Receiver Ready	*C	
14		RL	Remote Loopback		*C
15		IC	Incoming Call	*C	
16		SR	Signaling Rate Selector		*C
17	35	TT	Terminal Timing		*T
18		TM	Test Mode	*C	
19		SG	Signal Ground		
20		RC	Receive Common		
28		IS	Terminal in Service		*C
32		SS	Select Standby		*C
33		SQ	Signal Quality	*C	
34		NS	New Signal	*C	
36		SB	Standby Indicator		*C
37		SC	Send Common		

Signal Type: D = Data, C = Control, T = Timing
Note: On the DB37 connector that is commonly used for RS449;
Pins 3 and 21 are undefined. B = Return.

PC COM PORT-232

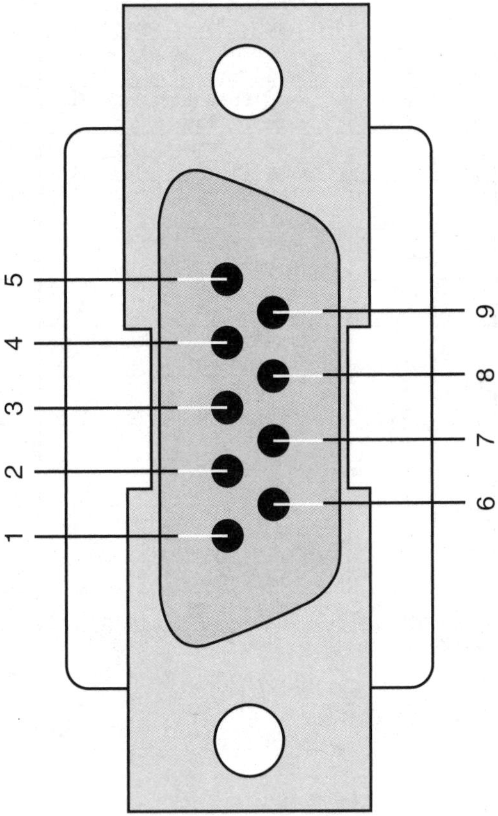

PC COM PORT-232 *(cont'd)*

Pin	Signal	Pin	Signal
1	Data Carrier Detect	6	Data Set Ready
2	Received Data	7	Request to Send
3	Transmitted Data	8	Clear to Send
4	Data Terminal Ready	9	Ring Indicator
5	Signal Ground		

EIA-449 SECONDARY INTERFACE

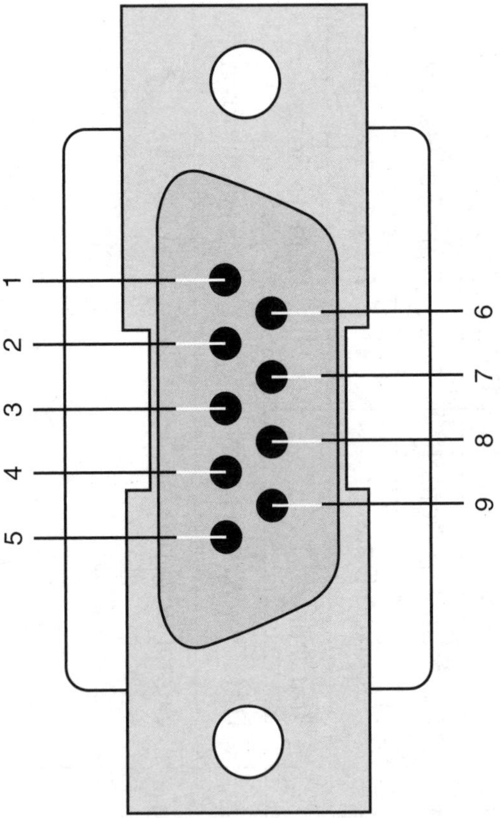

EIA-449 SECONDARY INTERFACE *(cont'd)*

Pin	Signal	Pin	Signal
1	Shield	6	Receive Common
2	Secopndary Receiver Ready	7	Secondary Request to Send
3	Secopndary Receiver Ready	8	Secondary Clear to Send
4	Secopndary Receiver Data	9	Send Common
5	Signal Ground		

CODE CONVERSION CHART

Hexadecimal	Decimal	ASCII Value	ASCII Character
$00	00	000	(null)
$01	01	001	☺
$02	02	002	☻
$03	03	003	♥
$04	04	004	♦
$05	05	005	♣
$06	06	006	♠
$07	07	007	(beep)
$08	08	008	■
$09	09	009	(tab)
$0A	10	010	(line feed)
$0B	11	011	(home)
$0C	12	012	(form feed)
$0D	13	013	(carriage return)
$0E	14	014	♫
$0F	15	015	☼
$10	16	016	►
$11	17	017	◄
$12	18	018	↕
$13	19	019	‼
$14	20	020	¶
$15	21	021	§
$16	22	022	_
$17	23	023	↨
$18	24	024	↑
$19	25	025	↓
$1A	26	026	→
$1B	27	027	←
$1C	28	028	(cursor right)
$1D	29	029	(cursor left)
$1E	20	030	(cursor up)
$1F	31	031	(cursor down)
$20	32	032	(space)
$21	33	033	!
$22	34	034	"
$23	35	035	#
$24	36	036	$
$25	37	037	%

CODE CONVERSION CHART *(cont'd)*

Hexadecimal	Decimal	ASCII Value	ASCII Character
$26	38	038	&
$27	39	039	'
$28	40	040	(
$29	41	041	)
$2A	42	042	*
$2B	43	043	+
$2C	44	044	,
$2D	45	045	-
$2E	46	046	.
$2F	47	047	/
$30	48	048	0
$31	49	049	1
$32	50	050	2
$33	51	051	3
$34	52	052	4
$35	53	053	5
$36	54	054	6
$37	55	055	7
$38	56	056	8
$39	57	057	9
$3A	58	058	:
$3B	59	059	;
$3C	60	060	<
$3D	61	061	=
$3E	62	062	>
$3F	63	063	?
$40	64	064	@
$41	65	065	A
$42	66	066	B
$43	67	067	C
$44	68	068	D
$45	69	069	E
$46	70	070	F
$47	71	071	G
$48	72	072	H
$49	73	073	I
$4A	74	074	J
$4B	75	075	K

CODE CONVERSION CHART *(cont'd)*

Hexadecimal	Decimal	ASCII Value	ASCII Character
$4C	76	076	L
$4D	77	077	M
$4E	78	078	N
$4F	79	079	O
$50	80	080	P
$51	81	081	Q
$52	82	082	R
$53	83	083	S
$54	84	084	T
$55	85	085	U
$56	86	086	V
$57	87	087	W
$58	88	088	X
$59	89	089	Y
$5A	90	090	Z
$5B	91	091	[
$5C	92	092	\
$5D	93	093	]
$5E	94	094	^
$5F	95	095	–
$60	96	096	'
$61	97	097	a
$62	98	098	b
$63	99	099	c
$64	100	100	d
$65	101	101	e
$66	102	102	f
$67	103	103	g
$68	104	104	h
$69	105	105	i
$6A	106	106	j
$6B	107	107	k
$6C	108	108	l
$6D	109	109	m
$6E	110	110	n
$6F	111	111	o
$70	112	112	p
$71	113	113	q

CODE CONVERSION CHART *(cont'd)*

Hexadecimal	Decimal	ASCII Value	ASCII Character
$72	114	114	r
$73	115	115	s
$74	116	116	t
$75	117	117	u
$76	118	118	v
$77	119	119	w
$78	120	120	x
$79	121	121	y
$7A	122	122	z
$7B	123	123	{
$7C	124	124	¦
$7D	125	125	}
$7E	126	126	~
$7F	127	127	⌂
$80	128	128	Ç
$81	129	129	ü
$82	130	130	é
$83	131	131	â
$84	132	132	ä
$85	133	133	à
$86	134	134	å
$87	135	135	ç
$88	136	136	ê
$89	137	137	ë
$8A	138	138	è
$8B	139	139	ï
$8C	140	140	î
$8D	141	141	ì
$8E	142	142	Ä
$8F	143	143	Å
$90	144	144	É
$91	145	145	æ
$92	146	146	Æ
$93	147	147	ô
$94	148	148	ö
$95	149	149	ò
$96	150	150	û
$97	151	151	ù

CODE CONVERSION CHART *(cont'd)*

Hexadecimal	Decimal	ASCII Value	ASCII Character
$98	152	152	ÿ
$99	153	153	Ö
$9A	154	154	Ü
$9B	155	155	¢
$9C	156	156	£
$9D	157	157	¥
$9E	158	158	P_t
$9F	159	159	*f*
$A0	160	160	á
$A1	161	161	í
$A2	162	162	ó
$A3	163	163	ú
$A4	164	164	ñ
$A5	165	165	Ñ
$A6	166	166	ª
$A7	167	167	º
$A8	168	168	¿
$A9	169	169	⌐
$AA	170	170	¬
$AB	171	171	½
$AC	172	172	¼
$AD	173	173	¡
$AE	174	174	«
$AF	175	175	»
$B0	176	176	░
$B1	177	177	▒
$B2	178	178	▓
$B3	179	179	│
$B4	180	180	┤
$B5	181	181	╡
$B6	182	182	╢
$B7	183	183	╖
$B8	184	184	╕
$B9	185	185	╣
$BA	186	186	║
$BB	187	187	╗
$BC	188	188	╝
$BD	189	189	╜

CODE CONVERSION CHART *(cont'd)*

Hexadecimal	Decimal	ASCII Value	ASCII Character
$BE	190	190	╛
$BF	191	191	╗
$C0	192	192	L
$C1	193	193	┴
$C2	194	194	┬
$C3	195	195	├
$C4	196	196	—
$C5	197	197	+
$C6	198	198	╞
$C7	199	199	╟
$C8	200	200	╚
$C9	201	201	╔
$CA	202	202	╩
$CB	203	203	╦
$CC	204	204	╠
$CD	205	205	=
$CE	206	206	╬
$CF	207	207	╧
$D0	208	208	╨
$D1	209	209	╤
$D2	210	210	╥
$D3	211	211	╙
$D4	212	212	╘
$D5	213	213	╒
$D6	214	214	╓
$D7	215	215	╫
$D8	216	216	╪
$D9	217	217	┘
$DA	218	218	┌
$DB	219	219	█
$DC	220	220	▄
$DD	221	221	▌
$DE	222	222	▐
$DF	223	223	▀
$E0	224	224	α
$E1	225	225	β
$E2	226	226	Γ
$E3	227	227	π

CODE CONVERSION CHART *(cont'd)*

Hexadecimal	Decimal	ASCII Value	ASCII Character
$E4	228	228	Σ
$E5	229	229	σ
$E6	230	230	μ
$E7	231	231	τ
$E8	232	232	Φ
$E9	233	233	Θ
$EA	234	234	Ω
$EB	235	235	δ
$EC	236	236	∞
$ED	237	237	ϕ
$EE	238	238	ϵ
$EF	239	239	∩
$F0	240	240	≡
$F1	241	241	±
$F2	242	242	≥
$F3	243	243	≤
$F4	244	244	⌠
$F5	245	245	⌡
$F6	246	246	÷
$F7	247	247	≈
$F8	248	248	°
$F9	249	249	·
$FA	250	250	·
$FB	251	251	√
$FC	252	252	η
$FD	253	253	^{2}
$FE	254	254	■
$FF	255	255	(blank 'FF')

COMPUTER TROUBLE-SHOOTING

Problem: Computer Starts Correctly, Then Stops

Possible Cause	Action
Computer is processing an endless loop or has stopped.	Reset the computer
Transient from power line.	Run original copy of the software again.
Bad copy of software.	Try a backup copy, or reinstall the software.
Problem in one of the peripherals or removable circuit cards.	Strip down to minimum working system. Remove as many circuit cards as possible (power off).
Hardware problem in main computer.	Try diagnostic programs.
Low voltage from power supply.	Measure output voltages.
Bad filtering noise from power supply.	Check for noise at each output.
No Clock signal to CPU.	Check with logic probe
No Reset signal to CPU.	Check with logic probe
Problem with data or address lines.	Use logic probe, diagnostics.
Faulty CPU IC.	Replace.

Computer Makes Intermittent, Random Errors

Possible Cause	Action
Transient voltages.	Run original copy of the software again.
Bad copy of software.	Try a backup copy or reinstall software.
Environmental problem.	Fault related to time of day, temperature, or nearby equipment switching on or off.
Bad memory IC.	Run diagnostic software.
Bad filtering, noise from power supply.	Check for noise at each output.
Faulty CPU IC.	Replace.

COMPUTER TROUBLE-SHOOTING *(cont'd)*

Computer Appears Dead

Possible Cause	Action
Problem in one of the peripherals or removable circuit cards.	Strip down to minimum working system. Remove as many circuit cards as possible, with power off.
Power supply problem.	Measure output voltages
No clock signal to CPU.	Check with logic probe.
No reset signal to CPU.	Check with logic probe.
Problem with data or address lines.	Logic probe, no-op tester.
Faulty CPU IC.	Replace.

Power On, Garbage On Screen

Possible Cause	Action
Transient from power line.	Reset. Turn computer off, wait 60 seconds, turn on again
Problem in one of the peripherals or removable circuit cards.	Strip down to minimum working system. Remove as many circuit cards as possible (power off). Try reset again.
Power supply problem.	Measure output voltage.
No clock signal to CPU.	Check with logic probe.
No reset signal to CPU.	Check with logic probe.
Problem with data or address lines.	Logic probe.
Problem with video driver circuits	Inspect, test.
Faulty CPU IC.	Replace.

Keyboard No Response

Possible Cause	Action
Bad cable or connector.	Clean, tighten connections.
Faulty keyboard decoder IC.	Ensure continuity through each wire of the cable. Substitute cable.

COMPUTER TROUBLE-SHOOTING *(cont'd)*

Keyboard No Response

Possible Cause	Action
Bad cable or connector.	Clean, tighten connections.
Faulty keyboard decoder IC.	Ensure continuity through each wire
of the cable. Substitute cable.	

Keyboard—Intermittent Problems

Possible Cause	Action
Bad cable or connector.	Clean, tighten connections. Ensure continuity through each wire of the cable. Substitute cable.
Faulty keyboard decoder IC.	Check, replace.

Disk Drive Will Not Operate

Possible Cause	Action
Jammed disk.	Try to remove the disk.
Loose part jamming the mechanism.	Open case, inspect.
Broken drive belt.	Inspect, check belt.
No power to drive motor.	Drive motor connections.
Bad cable or connectors.	Clean, tighten connections. Ensure continuity through each wire in the cable. Substitute cable.
Fault in interface circuits.	Substitute another interface card.
Fault in control circuits in the drive itself.	Replace the suspect drive. Swap drives to verify.
Fault in control circuits.	Ensure drive motor is on and signal is present.

COMPUTER TROUBLE-SHOOTING *(cont'd)*

Disk Drive—Intermittent Read or Write Failure

Possible Cause	Action
Dirty read/write head.	Clean the head, try the disk again.
Hardware or software error indicated by error statement.	Check manual.
Bad disk	Inspect for mechanical damage. Substitute another disk with a copy of the same program or file. Use a utility to check the suspect disk for bad sectors. Substitute a backup copy of the same program or file.
Software program.	Look for a pattern. Try other software.
Bad cable or connectors.	Clean, tighten connections. Ensure continuity through each wire of the cable. Substitute cable.
Fault in interface circuits.	Substitute another interface card.
Fault in control circuits in the drive itself.	Substitute the suspect drive. Swap drives to verify.
Environmental problems.	Fault related to time of day, temperature, nearby equipment switching on or off, etc.
Power-supply problem.	Noise, bad ground. Correct.
Incorrect motor speed.	Adjust motor speed.
Drive out of alignment.	Alignment procedure.
Fault in control circuits.	Check for correct signals through the cable.

Disk Drive—Cannot Insert or Remove

Possible Cause	Action
Mechanical parts slightly out of alignment.	Open and close door a few times, try again.
Head remains loaded.	Reset.
Mechanical eject—bent part.	Open case, inspect.
Auto eject—bent part.	Open case, inspect. Check for bad solenoid.
Faulty motor-driven ejector.	Inspect ejector.

COMPUTER TROUBLE-SHOOTING *(cont'd)*

Disk Drive Will Not Read

Possible Cause	Action
Dirty Read/Write head	Clean head, try original disk again.
Hardware or software error indicated by error statement.	Check manual.
Bad disk	Inspect for mechanical damage. Substitute another disk with a copy of the same program or file. Use a utility to check the suspect disk for bad sectors.
Software problem.	Look for a pattern. Try other software.
Bad cable or connectors.	Clean, tighten connections. Ensure continuity through each wire in cable. Substitute cable.
Fault in interface circuits.	Substitute another interface card.
Fault in control circuits in the drive itself.	Substitute the suspect drive. Swap drives to verify.
Worn head-load pad.	Inspect load pad.
Drive select problem.	Check for drive-select signal through the cable.
Faulty read circuits.	Check for read-data signals through the cable.
Drive out of alignment.	Alignment procedure.

Disk Drive—Damaged Files on Disk

Possible Cause	Action
Recoverable fault on disk.	Recovery procedures.

Disk Drive—Head Oscillates at Track 0

Possible Cause	Action
Faulty Tract-zero detector.	Test detector.
Faulty Tract-zero logic.	Check circuit.

COMPUTER TROUBLE-SHOOTING *(cont'd)*

Disk Drive—Drive Turns Slowly, Makes Noise

Possible Cause	Action
Faulty or jammed disk.	Try to remove disk.
Loose part jamming the mechanism.	Open case, inspect.
Loose, slipping drive belt.	Inspect, check tension.
Inadequate power to drive motor.	Ensure correct power is available. Check voltages through the cable.
Bad cable or connectors.	Clean, tighten connections. Ensure continuity through each wire in the cable. Substitute cable.
Motor speed needs adjustment.	Run motor-speed test software.
Mechanical problem—drive motor and spindle.	Raw data test. Inspect drive motor and spindle.

Disk Drive Will Not Write

Possible Cause	Action
Hardware or software error indicated by error statement.	Check the manual.
Disk is write-protected.	Inspect write-protect notch.
Bad disk	Inspect for mechanical damage. Substitute another disk and try to write again. Use a utility to check the suspect disk for bad sectors.
Software problem.	Look for a pattern. Try other software. Can the drive write while under control of these other programs?
Bad cable or connectors.	Clean, tighten connections. Ensure continuity through each wire of the cable. Substitute cable.
Fault in interface circuits.	Substitute another interface card.
Fault in control circuits in the drive itself.	Substitute the suspect drive. Swap drives to verify.
Drive out of alignment.	Alignment procedure.
Faulty write-protect circuit.	Inspect components, test the circuit.
Faulty write circuits.	Inspect components, test the circuit.
Fault in control circuits.	Check for write-gate and write-data signals through the cable.

COMPUTER TROUBLE-SHOOTING *(cont'd)*

Disk Drive—Head Will Not Step In or Out

Possible Cause	Action
Fault in stepper logic or stepper motor.	Inspect mechanism, test circuit.

Disk Drive—Cannot Read

Possible Cause	Action
Drive is out of alignment.	Alignment procedure.

Hard Drive—Damaged Files

Possible Cause	Action
Recoverable fault on disk.	Recovery procedures.

Hard Drive—Intermittent Read or Write Failures

Possible Cause	Action
Identifiable error.	Look up error statement. Check self-diagnostics.
Break in cable, bad connector.	Continuity through cable.
Bad connection on card connector.	Check card connections.
Problem on interface card.	Substitute card.
Problem with one circuit on card with duplicate circuits.	Connect drive to second circuit.
Low voltage or electronic noise.	Check power-supply output voltages.
Internal drive causes overheating.	Install fan.
Internal drive overloads power supply.	Install more powerful supply.
Worn bearings.	Check for noise from drive.
Worn Read/Write heads.	Bit-shift test.
Weak Read/Write output.	Read circuits. Write circuits.
Faulty index detector circuits.	Check index detector.

COMPUTER TROUBLE-SHOOTING *(cont'd)*

Hard Drive Head Oscillates at Track 000

Possible Cause	Action
Faulty Track 000 circuits.	Track 000 system.

Hard Drive—Unusual Noise From Drive

Possible Cause	Action
Worn bearings, damaged head.	Return to manufacturer.

Hard Drive Head Will Not Step In or Out

Possible Cause	Action
Identifiable error.	Look up error statement. Check self-diagnostics.
Break in cable; bad connector.	Check continuity through cable. Check signals through cable: "Step," "Direction."
Bad connection at card connector.	Check card connections.
Problem on interface card.	Substitute card.
Problem with one circuit on card with duplicate circuits.	Connect drive to second circuit.
Faulty signals to stepper motor.	Check signals to stepper motor.
Faulty signals to voice coil.	Check signals to voice coil.
Mechanical problem.	Return drive to manufacturer.

COMPUTER TROUBLE-SHOOTING *(cont'd)*

Hard Drive Operates, But Will Not Write

Possible Cause	Action
Identifiable error.	Look up error statement. Check self-diagnostics.
Break in cable; bad connector.	Check continuity through cable. Check signals through cable: "Write Gate," "Write Data," "Write Enable."
Bad connection at card connector.	Check card connections.
Problem on interface card.	Substitute card.
Problem with one circuit on card with duplicate circuits.	Connect drive to second circuit.
Worn Read/Write heads	Bit shift test.
Weak or absent write signals.	Check write circuits.

Hard Drive Will Not Read

Possible Cause	Action
Identifiable error.	Look up error statement. Check self-diagnostics.
Break in cable, bad connector.	Check continuity through cable. Check signals through cable: "Read Gate," "Read Data," "Read Clock," "Read Enable."
Bad connection at card connector.	Check card connections.
Problem on interface card.	Substitute card.
Problem with one circuit on card with duplicate circuits.	Connect drive to second circuit.
Worn Read/Write heads.	Bit shift test.
Weak or absent read signals.	Check read circuits.

PRINTER TROUBLE-SHOOTING

Laser Printer—No Printing; White Page

Possible Cause	Action
No DC bias on developing cylinder.	Image development check.
No ground to photosensitive drum.	Remove cartridge and check connections.
Drum not turning.	Check movement of drum.
Cartridge out of toner.	Replace cartridge.
Laser beam is blocked.	Inspect beam path.
No transfer corona.	Inspect transfer corona wire. Check corona voltage.

Laser Printer—Black Page

Possible Cause	Action
No primary corona.	Inspect primary corona wire. Check for corona voltage.
Laser permanently on.	Check control circuits.
Damaged beam detector.	Inspect beam detector.

Laser Printer—Horizontal Black Lines

Possible Cause	Action
No beam detect signal.	Inspect detector.

Laser Printer—Black Page with White Horizontal Lines

Possible Cause	Action
No beam detect signal.	Beam detector.

PRINTER TROUBLE-SHOOTING *(cont'd)*

Laser Printer—Random Dot Patterns

Possible Cause	Action
Incorrect signals from host computer. Faulty control circuits.	Try self-test. Printer CPU.

Laser Printer—Smeared Print

Possible Cause	Action
No fusing action.	Heating action of rollers. Fusing lamp. Fusing temperature controls.
Static charge eliminator teeth bent.	Inspect teeth.
Incorrect type of paper.	Use correct paper.

Laser Printer—Repeating Marks on Paper

Possible Cause	Action
Dirt or mark on a roller.	Measure distance, calculate diameter, inspect and clean roller.

Laser Printer—Very Light Print

Possible Cause	Action
Cartridge is out of toner.	Remove cartridge, shake, and replace.
No transfer corona.	Inspect transfer corona wire. Check for corona voltage.
No DC bias on developing cylinder.	Check for DC bias.

PRINTER TROUBLE-SHOOTING *(cont'd)*

Laser Printer—Left or Right Side of Page Is Blank

Possible Cause	Action
Laser scanning to one side.	Check large mirror over photo-sensitive drum.

Laser Printer—Distorted Print

Possible Cause	Action
Drive rollers worn or dirty. Drive gears worn or clogged.	Inspect, replace. Inspect, replace.

Laser Printer—Light Print

Possible Cause	Action
Cartridge running out of toner.	Remove cartridge, rock it and replace it.
Faulty cartridge sensitivity switches. Erase lamps not working.	Check switches. Check lamps.

Laser Printer—Dark Blotches

Possible Cause	Action
No primary corona.	Inspect primary corona wire. Check for corona voltage.

MONITOR TROUBLE-SHOOTING

Laser Printer—Vertical White Streaks

Possible Cause	Action
Dirty mirror above photosensitive drum.	Inspect, clean.
Dirty transfer corona assembly.	Inspect, clean.
Cartridge running out of toner.	Remove cartridge, shake, and replace.

Monitor—Shortened Picture or White Line Across Screen

Possible Cause	Action
Vertical size adjustment.	Adjust.
Weak vertical output.	Check vertical oscillator and vertical output circuits.

Monitor—Images Have Colored Shading on One or Two Sides

Possible Causes	Action
Monitor out of alignment.	Alignment procedure.

Monitor—Faint Lines Behind the Picture

Possible Cause	Action
Black level setting	Adjust.

MONITOR TROUBLE-SHOOTING *(cont'd)*

Monitor—Picture Stretched Sideways, Black Lines Visible

Possible Cause	Action
Picture out of sinc.	Horizontal oscillator and horizontal output sections.

Monitor—Narrow Picture

Possible Cause	Action
Weak horizontal output.	Horizontal oscillator and horizontal output circuits.

Monitor—Picture Rolls Side-to-Side

Possible Cause	Action
Horizontal hold adjustment.	Adjust.
Horizontal sync and horizontal output circuits.	Check circuits.

Monitor—Picture Rolls Up or Down

Possible Cause	Action
Vertical hold adjustment.	Adjust.
Vertical sync and vertical output circuits.	Check circuits.

MONITOR TROUBLE-SHOOTING *(cont'd)*

Monitor—Screen Is Dark

Possible Cause	Action
No power to monitor. Short circuit causing "foldback." No electron beam.	AC outlet, plug, fuse, circuit breaker. Use variac. Picture tube, high-voltage circuits, scanning circuits.

Monitor—Screen Lights, but No Image Displayed

Possible Cause	Action
No modulation to electron beam.	Video output and sync separator circuits.

Monitor—Sections of Picture Waver

Possible Cause	Action
Bad filter capacitor in power supply.	Check capacitor.

Monitor—Picture Shrinks

Possible Cause	Action
Faulty power supply. Beam not deflected outward.	Check output voltages. Horizontal and vertical outputs.

MODEM TROUBLE-SHOOTING

Modem—No Ready Indicator

Possible Cause	Action
No power to modem.	See if modem is plugged in.
Modem set to incorrect mode.	Check mode setting. (One modem should be set to originate, other to answer.)
Fault in modem.	Try self-test.
Problem with RS-232C interface.	Check for correct signals through RS-232C cable.
Mechanical damage to modem circuit board.	Inspect circuit board.
Faulty modem IC.	Check, replace.
Fault in serial interface circuits in computer.	Remove, replace circuits (computers with removable circuit cards). Inspect, replace UART IC.

Modem—Intermittent Problem

Possible Cause	Action
Fault inside modem.	Try modem self-test.
Noisy phone line (long-distance section).	Listen to line (line bad only on one call), hang up, call back.
Noisy phone line (local section).	Listen to line (line bad on all calls), call phone company.
Bad cable or connectors.	Clean, tighten connections. Ensure continuity through each wire in the cable. Substitute cable.
Faulty modem IC.	Check, replace.
Fault in serial interface circuits in computer	Remove, replace circuits (computers with removable circuit cards). Inspect, replace UART IC.

INTERNET TERMS

Bandwidth. The maximum amount of information that can be transmitted at any given time. A 56k leased line connection, for example, has 56k of bandwidth.

Client. A program that is run by users on their machine. It issues requests to a server, which is generally located on another system. This takes a big work-load off of the server program, so that it can process client requests more efficiently. This also makes the system appear very fast.

CSLIP (Compressed SLIP). SLIP with compression for a more efficient connection. See SLIP.

ECPA (Electronic Communications Privacy Act). A law passed a few years back that says that all electronic mail cannot be read by the people running the system. Its main concrete achievement seems to have been the placing of a notice on all bulletin board systems (BBSs) saying that there is no private mail function on their systems, despite the continued existence of same on the menu.

Flame. An insulting message, normally with little real content. A Flame War is a seemingly endless exchange of such messages.

FTP. File Transfer Protocol. This refers to a protocol describing the way files can be transferred over a TCP/IP network, such as the Internet. The program used to implement this protocol is also called FTP. Normally, a FTP program is included with basic networking software, and little needs to be done to make it work on a system.

HTML. (HyperText Markup Language). This is the scheme used to design World Wide Web pages. There are numerous tools that can help you write HTML with reasonable efficiency.

HTTP. (HyperText Transfer Protocol). This is the protocol used for information transmitted over the World Wide Web (WWW).

InterNIC. The government-funded service, run by a company called Network Solutions, that parcels out IP addresses and domain names. Complaints about slow service have been heard quite loudly, and the entire system is being revised.

IRC (Internet Relay Chart). This is a direct interactive way for people to hold conversations using the computer.

Java. This is a programming language created by Sun. It was originally designed for the development of proprietary information appliances, but is now being widely used on the web.

Microsoft Internet Explorer. Microsoft's "browser" software—software that make surfing the net easy and painless.

MOSAIC. The first truly effective web browser.

MUD or Multi-User Domain. These are the 'virtual worlds' created by serious Internet users. They were originally developed by game-players, but are now being used for commercial purposes.

INTERNET TERMS *(cont'd)*

Netscape. The Web Browser that made the Web a hit. Now competing with Microsoft Explorer.

News, also called **USENET.** This is a messaging system that is one of the most famous and popular parts of the net. It functions as an electronic bulletin board. Each of the thousands of newsgroups are devoted to a particular topic.

PPP. Point-to-Point Protocol. A newer and supposedly better way to connect your site to the Internet via a single serial line. Windows95 has greatly expanded its popularity, since it supports PPP instead of the older SLIP. See SLIP.

RFC. Request for Comment. This is an informal system for proposing Internet standards. The technical people who work on the Internet upload RFCs to the NIC, where they are given a number and published. Many of them are later adopted as Internet standards.

Search Engine. As the World Wide Web has grown bigger and bigger, programs have been created that wander the web, looking for resources of interest. They then put them in enormous keyword dictionaries and let you search for what you'd like to find. Two search engines are Digital's Altavista [***http://altavista.digital.com***], and Lycos [***http://www.lycos.com/***]. See also Yahoo, which is really a directory, not a search engine.

Server. A program running on a remote system that provides information to a client. See Client for a detailed explanation.

SLIP. Serial In-Line Protocol. This is one of several ways to attach a computer to the Internet via a simple (and cheap) modem connection. See the earlier discussion on connecting your system to the Internet for additional information.

TCP/IP, Transmission Control Protocol/Internet Protocol. The protocol used to send information through the Internet.

TELNET. Telnet is a program that lets you remotely log in to any other system on the Internet (assuming you have access).

WWW (World Wide Web). This is probably the best Internet browsing system—certainly the most fun one to use.

Yahoo. Yahoo is a comprehensive directory of WWW resources. Has now become ubiquitous.

HTML TERMS

Absolute URI: A URI in absolute form.

Anchor: One of two ends of a hyperlink; typically, a phrase marked as an *A* element.

Base URI: An absolute URI used in combination with a relative URI to determine another absolute URI.

Character: A unit of information, a letter or a digit.

Character Encoding Scheme: A function whose domain is the set of sequences of characters from a character repertoire; that is, a sequence of octets and a character encoding scheme determines a sequence of characters.

Character Repertoire: A finite set of characters; e.g. the range of a coded character set.

Code Position: An integer. A coded character set and a code position from its domain determine a character.

Conforming HTML User Agent: A user agent that conforms to this specification in its processing of the Internet Media Type 'text/html'.

Data Character: Characters other than markup, which make up the content of elements.

Document Character Set: A coded character set whose range includes all characters used in a document.

DTD: Document type definition.

Element: A component of the hierarchial structure defined by a document type definition; it is identified in a document instance by descriptive markup, usually a start-tag and end-tag.

End-Tag: Descriptive markup that identifies the end of an element.

Entity: Data with an associated notation or interpretation; for example, a sequence of octets associated with an Inernet Media Type.

Fragment Identifier: The portion of an HREF attribute value following the '#' character which modifies the presentation of the destination of a hyperlink.

Form Data Set: A sequence of name/value pairs; the names are given by an HTML document and the values are given by a user.

HTML Document: An SGML document conforming to this document type I definition.

Hyperlink: A relationship between two anchors, called the head and the tail. The link goes from the tail to the head. The head and tail are also known as destination and source, respectively.

HTML TERMS *(cont'd)*

Markup: Syntactically delimited characters added to the data of a document to represent its structure. There are four different kinds of markup: descriptive markup (tags), references, markup declarations, and processing instructions.

Media Type: An internet Media Type.

Message Entity: A head and body. The head is a collection of name/value fields, and the body is a sequence of octets. The head defines the content transfer encoding of the body.

Must: Documents or user agents in conflict with this statement are not conforming.

Numeric Character Reference: Markup that refers to its character by its code position in the document character set.

SGML Document: A sequence of characters organized physically as a set of entities and logically into a hierarchy of elements. An SGML document consists of data characters and markup; the markup describes the structure of the information and an instance of that structure.

Start-Tag: Descriptive markup that identifies the start of an element and specifies its generic identifier and attributes.

Tag: Markup that delimits an element. A tag includes a name which refers to an element declaration in the DTD, and may include attributes.

Text Entity: A finite sequence of characters. A text entity typically takes the form of a sequence of octets with some associated character encoding scheme, transmitted over the network or stored in a file.

Typical: Typical processing is described for any elements. This is not a mandatory part of the specification but is given as guidance for designers and to help explain the uses for which the elements were intended.

URI: A Uniform Resource Identifier is a formatted string that serves as an identifier for a resource, typically on the Internet. URIs are used in HTML to identify the anchors of hyperlinks. URIs in common practice include Uniform Resource Locators (URLs) and Relative URLs.

User Agent: A component of a distributed system that presents an interface and processes requests on behalf of a user; for example, a WWW browser or mail user agent.

WWW: The World Wide Web is a hypertext-based, distributed information system created by researchers at CERN in Switzerland. http://www.w3.org/

HTML DOCUMENT STRUCTURE

```
<!DOCTYPE HTML PUBLIC  "-//IETF//DTD HTML 2.0//EN">
<HTML>
<!--comments-->
<HEAD>
<TITLE>Structural Example</TITLE>
</HEAD><BODY>
<H1>First Header</H1>
<P>This is a paragraph in the example HTML file.  Keep in mind that a title does
not appear in the document text, but that the header (defined in H1) does.</P>
<OL>
<LI>First item is an ordered list.
<LI>Second item is an ordered list.
 <UL COMPACT>
 <LI>Note that lists can be nested;
 <LI>Whitespace may be used to assist in reading the HTML source.
 </UL>
<LI>Third item in an ordered list.
</OL>
<P>This is an additional paragraph.  Technically, end tags are not required for
paragraphs, although they are allowed.  You can include character highlighting in
a paragraph.  <EM>This sentence of the paragraph is emphasized. </EM> Note
that the &lt;/P&gt; end tag is has been omitted.
<P>
<IMG SRC="triangle.xbm" alt="Warning: ">
Be sure to read these <B>bold instructions</B>.
</BODY></HTML>
```

SYMBOLS

URL	URL of an external file (or just file name if in the same directory)
?	Arbitrary number (i.e. <H?> means <H1>, <H2>, <H3>, etc.)
%	Arbitrary percentage (i.e. <HR WIDTH="%" means <HR WIDTH= "50%">, etc.)
***	Arbitrary text (i.e. ALT="****" means fill in with text)
$$$$$$	Arbitrary hex (i.e. BGCOLOR="$$$$$$" means BGCOLOR= "#00ff1C", etc.
:::	Arbitrary date (i.e. DATETIME=":::" means "1994-11-05T08:15:30" etc.)
@	Email address (i.e. "mailto:@" means "mailto:jdoe@isp.com", etc.)
,,,	Comma-delimited (i.e. COORDS=",,," means COORDS="0,0,50,50", etc.)
I	Alternatives (i.e. ALIGN=LEFTI RIGHTI CENTER means pick one of these)

GENERAL

Document Type	<HTML></HTML>	(beginning and end of file)
Title	<TITLE></TITLE>	(must be in header)
Header	<HEAD></HEAD>	(descriptive info, such as title)
Body	<BODY></BODY>	(bulk of the page)

STRUCTURAL DEFINITION

Heading	<H?></H?>	(spec. defines 6 levels)
Align Heading	<H? ALIGN=LEFTI RIGHTI CENTERI JUSTIFY></DIV>	
Division	<DIV></DIV>	
Align Division	<DIV ALIGN=LEFTI RIGHTI CENTERI JUSTIFY></DIV>	
Defined Content	<SPAN></SPAN>	
Block Quote	<BLOCKQUOTE></BLOCKQUOTE>	(usually indented)

STRUCTURAL DEFINITION *(cont'd)*

Quote	<Q></Q>	(for short quotations)
Citation	<Q CITE="URL"></Q>	
Emphasis	<EM></EM>	(usually displayed as italic)
Strong Emphasis	<STRONG></STRONG>	(usually displayed as bold)
Citation	<CITE></CITE>	(usually italics)
Code	<CODE></CODE>	(for source code listings)
Sample Output	<SAMP></SAMP>	
Keyboard Input	<KBD></KBD>	
Variable	<VAR></VAR>	
Definition	<DFN></DFN>	
Author's Address	<ADDRESS></ADDRESS>	
Large Font Size	<BIG></BIG>	
Small Font Size	<SMALL></SMALL>	
Insert	<INS></INS>	
Time of Change	<INS DATETIME=":::"></INS>	
Comments	<INS CITE="URL"></INS>	
Delete	<DEL></DEL>	
Time of Change	<DEL DATETIME=":::"></DEL>	
Comments	<DEL CITE="URL"></DEL>	
Acronym	<ACRONYM></ACRONYM>	
Abbreviation	<ABBR></ABBR>	

PRESENTATION FORMATTING

Bold	<B></B>
Italic	<I></I>
Underline	<U></U>
Strikeout	<STRIKE></STRIKE>

PRESENTATION FORMATTING *(cont'd)*

Strikeout	<S></S>	
Subscript		
Superscript		
Typewriter	<TT></TT>	(displays in a monospaced font)
Preformatted	<PRE></PRE>	(displays text spacing as-is)
Width	<PRE WIDTH=?></PRE>	(in characters)
Center	<CENTER></CENTER>	(for both text and images)
Blinking	<BLINK></BLINK>	(the most derided tag ever)
Font Size	<FONT SIZE=?></FONT>	(ranges from 1-7)
Change Font Size	<FONT SIZE="+I-?"></FONT>	
Font Color	<FONT COLOR="#$$$$$$"></FONT>	
Select Font	<FONT FACE="****"></FONT>	
Point Size	<FONT POINT-SIZE=?></FONT>	
Weight	<FONT WEIGHT=?></FONT>	
Base Font Size	<BASEFONT SIZE=?>	<from 1-7; default is 3)
Marquee	<MARQUEE></MARQUEE>	

DIVIDERS

Paragraph	<P></P>	(closing tag often unnecessary)
Align Text	<P ALIGN=LEFTI CENTERI RIGHT></P>	
Justify Text	<P ALIGN=JUSTIFY></P>	
Line Break	 	(a single carriage return)
Clear Textwrap	<BR CLEAR=LEFTI RIGHTI ALL>	
Horizontal Rule	<HR>	
Alignment	<HR ALIGN=LEFTI RIGHTI CENTER>	
Thickness	<HR SIZE=?>	(in pixels)

DIVIDERS *(cont'd)*

Width	<HR WIDTH=?>	(in pixels)
Width Percent	<HR WIDTH="%">	(as a percentage of page width)
Solid Line	<HR NOSHADE>	
No Break	<NOBR></NOBR>	(prevents line breaks)
Word Break	<WBR>	(where to break a line if needed)

LISTS (lists can be nested)

Unordered Lists	<UL><LI></UL>	(<LI> before each list item)
Compact	<UL COMPACT></UL>	
Bullet Type	<UL TYPE=DISC| CIRCLE| SQUARE>	(for the whole list)
	<LI TYPE=DISC| CIRCLE| SQUARE>	(this & subsequent)
Ordered List	<OL><LI></OL>	(<LI> before each list item)
Compact	<OL COMPACT></OL>	
Numbering Type	<OL TYPE=A|a|I|i|1>	(for the whole list)
	<LI TYPE=A|a|I|i|1>	(this & subsequent)
Starting Number	<OL START=?>	(for the whole list)
	<LI VALUE=?>	(this & subsequent)
Definition List	<DL><DT><DD></DL>	(<DT>=term, <DD>=definition)
Compact	<DL COMPACT></DL>	
Menu List	<MENU><LI></MENU>	(<LI> before each list item)
Compact	<MENU COMPACT></MENU>	
Directory List	<DIR><LI></DIR>	(<LI> before each list item)
Compact	<DIR COMPACT></DIR>	

BACKGROUNDS AND COLORS

Tiled Background — <BODY BACKGROUND="URL">

Watermark — <BODY BG PROPERTIES="FIXED">

Background Color — <BODY BGCOLOR="#$$$$$$">(order is red/green/blue)

Text Color — <BODY TEXT="#$$$$$$”>

Link Color — <BODY LINK="#$$$$$$”>

Visited Link — <BODY VLINK="#$$$$$$”>

Active Link — <BODY ALINK="#$$$$$$”>

SPECIAL CHARACTERS (must be lowercase)

Special Character — &#?;

< — <

> — >

& — &

“ — "

Registered TM — &#174

Registered TM — ®

Copyright — ©

Copyright — ©

Non-Breaking Space —

FORMS

Define Form — <FORM ACTION="URL" METHOD=GET| POST></FORM>

File Upload — <FORM ENCTYPE="multipart/form-data"></FORM>

Input Field — <INPUT TYPE="TEXT| PASSWORD| CHECKBOX| RADIO| FILE| BUTTON| IMAGE| HIDDEN| SUBMIT| RESET">

Field Name — <INPUT NAME="****">

FORMS *(cont'd)*

Field Value	<INPUT VALUE="****">
Checked?	<INPUT CHECKED>
Field Size	<INPUT SIZE=?>
Max Length	<INPUT MAXLENGTH=?>
Button	<BUTTON></BUTTON>
Button Name	<BUTTON NAME="****"</BUTTON>
Button Type	<BUTTON TYPE="SUBMI TIRESETI BUTTON"> </BUTTON>
Default Value	<BUTTON VALUE="****"></BUTTON>
Label	<LABEL></LABEL>
Item Labelled	<LABEL FOR="****"></LABEL>
Selection List	<SELECT></SELECT>
Name of List	<SELECT NAME="****"></SELECT>
# of Options	<SELECT SIZE=?></SELECT>
Multiple Choice	<SELECT MULTIPLE>
Option	<OPTION>
Default Option	<OPTION SELECTED>
Option Value	<OPTION VALUE"****">
Option Group	<OPTGROUP LABEL"****"></OPTGROUP>
Input Box Size	<TEXTAREA ROWS=? COLS=?></TEXTAREA>
Name of Box	<TEXTAREA NAME="****"</TEXTAREA>
Wrap Text	<TEXTAREA WRAP=OFFI HARDI SOFT></TEXTAREA>
Group Elements	<FIELDSET></FIELDSET>
Legend	<LEGEND></LEGEND> (caption for fieldsets)
Alignment	<LEGEND ALIGN="TOPI BOTTOMI LEFTI RIGHT"> </LEGEND>

LINKS, GRAPHICS, AND SOUND

Specify Subject <A HREF="mailto:@?SUBJECT=***"></A>

Define Location <A NAME="****"></A>

Display Image <IMG SRC="URL">

Alignment <IMG SRC="URL" ALIGN=TOP| BOTTOM| MIDDLE| LEFT| RIGHT>

Alignment <IMG SRC="URL" ALIGN=TEXTTOP| ABSMIDDLE| BASELINE| ABSBOTTOM>

Alternate <IMG SRC="URL" ALT="****"> (if image not displayed)

Dimensions <IMG SRC="URL" WIDTH=? HEIGHT=?> (in pixels)
<IMG SRC="URL" WIDTH=% HEIGHT=%> (as %)

Border <IMG SRC="URL" BORDER=?> (in pixels)

Runaround Space <IMG SRC="URL" HSPACE=? VSPACE=?> (in pixels)

Low-Res Proxy <IMG SRC="URL" LOWSRC="URL">

Image Map <IMG SRC="URL" ISMAP> (requires a script)

Image Map <IMG SRC="URL" USEMAP="URL">

Movie Clip <IMG DYNSRC="****" START="****" LOOP=?>

Background Sound <BGSOUND SRC="****" LOOP=?| INFINITE?

Client-Side Map <MAP NAME="****"></MAP> (describes the map)

Map Section <AREA SHAPE="DEFAULT| RECT| CIRCLE| POLY" COORDS=" ,,," HREF="URL"|NOREF>

Client Pull <META HTTP-EQUIV="Refresh" CONTENT="?; URL=URL

Embed Object <EMBED SRC="URL"> (insert object into page)

Object Size EMBED SRC="URL" WIDTH=? HEIGHT=?

Object <OBJECT></OBJECT>

Parameters <PARAM>

Link Something <A HREF="URL"></A>

Link to Location <A HREF="URL#***"></A> (if in another document)
<A HREF="#***"></A> (if in current document)

Target Window <A HREF="URL" TARGET="****"></A>

Action on Click <A HREF="URL" ONCLICK="****"></A> (Javascript)

Link to Email <A HREF="mailto:@"></A>

CHAPTER 9
Glossary

GLOSSARY

10Base2: 10 Mbps, baseband, in 185 meter segments. The IEEE 802.3 substandard for ThinWire, coaxial, Ethernet.

10BaseT: 10 Mbps, baseband, over twisted pair. The IEEE 802.3 substandard for unshielded twisted pair Ethernet.

110 Type Block: A wire connecting block that terminates 100 to 300 pairs of wire. It has excellent electrical characteristics and relatively small foot print. It organizes pairs horizontally.

66-Type Block: A type of wire connecting block that is used for twisted pair cabling cross connections. It holds 25 pairs in one to four vertical columns.

80x86: Family of microprocessors made by Intel used in PC and clone computers.

A/D converter: Analog to digital converter.

Absorption: Loss of power in an optical fiber, resulting from conversion of optical power into heat and caused principally by impurities, such as transition metals and hydroxyl ions, and also by exposure to nuclear radiation.

Acceptance angle: the half-angle of the cone within which incident light is totally internally reflected by the fiber core. It is equal to arcsin (NA).

Acceptance Test: A test run on a host or network to determine its operation is satisfactory prior to acceptance by the purchaser or the purchaser's agent.

Access Charge: A charge, set by the FCC (Federal Communications Commission), for access to a local carrier by a user or long- distance supplier.

Access Code: Digits a user must enter to obtain access to a particular service, physical area or system.

Access Control Byte: Token ring field that holds the priority and reservation bits for a token or data packet.

Access (noun): Level of authority or point of presence a user has for reading data.

Access (verb): To connect, as in: to data in memory; to a peripheral (or another device) for use; to another host.

Access floor: A system of raised flooring that has completely removable and interchangeable floor panels. The floor panels are supported on adjustable pedestals or stringers (or both) to allow access to the area beneath.

GLOSSARY *(cont.)*

Access Line: A line or circuit that connects a customer site to a network switching center or local exchange. Also known as the local loop.

Access Method: (1) A routine that prepares data for transmission. (2) A way of connecting to a host or peripheral. Also known as access routine.

Access Time: The period between a request to store or read data and the completion of that storage or retrieval.

Accessible: Can be removed or exposed without damaging the building structure. Not permanently closed in by the structure or finish of the building.

Accessible, readily (readily accessible): Can be reached quickly, without climbing over obstacles or using ladders.

Active Device: A device with its own power source.

Active Hub: A device used to amplify transmission signals in certain network topologies. An active hub can be used to either add additional workstations to a network or to lengthen the cable distance between nodes (workstations and/or file servers) on a network.

Adapter: A connectivity device that links different parts of one or more subsystems, systems, hosts or networks. For example: a network interface card may also be called a network adapter.

Add-on-board: An optional circuit board that conveniently modifies or enhances a personal computer's capabilities. See also Memory Board, Network Interface Board, Network Interface Card.

Address (computer): The designation of a particular word location in a computer's memory or other data register.

Address: An identified name for a logical or physical item that can be used to connect to that item.

ADP: Automated Data Processing, Automatic Data Processing, Administrative Data Processing, Advanced Data Processing.

Aerial Cable: Telecommunications cable installed on aerial supporting structures such as poles, sides of buildings, and other structures.

Aggregate: A masonry substance that is poured into place, then sets and hardens, as concrete.

AI: Artificial Intelligence.

AIA: American Institute of Architects.

Air Handling Plenum: A designated area, closed or open, used for environmental air circulation.

GLOSSARY *(cont.)*

Algorithm: A well-defined set of rules that solve a problem in a finite number of steps. For example, a full arithmetic procedure for determining retransmission wait time.

Alphanumeric: A character set that contains letters, numerals (digits), and other characters such as punctuation. Also used to identify one of such characters.

Alpha Test: The first test of newly developed software or hardware (usually in a laboratory). See also beta test.

Alternate Route: A secondary communications path used to reach a destination if the primary path is unavailable.

Alternate Routing: A way of completing connections that use another path when the previous circuit is unavailable, busy or out of service.

Alternate Use: The ability to switch communications facilities from one type to another, i.e., voice to data, etc.

Alternating current (AC): Electrical current which reverses direction repeatedly and rapidly. The change in current is due to a change in voltage which occurs at the same frequency.

AM: Amplitude modulation.

Ambient temperature: The temperature of the surroundings.

American National Standards Institute (ANSI): A private organization that coordinates some US standards setting. It also approves some US standards that are often called ANSI standards. ANSI also represents the United States to the International Standards Organization. See also: International Standards Organization.

American Standard Code for Information Interchange (ASCII): A standard character set that (typically) assigns a 7-bit sequence to each letter, number, and selected control character. Erroneously used now to refer to (8-bit) Extended ASCII. The other major encoding standard is EBCDIC.

American Wire Gauge (AWG): A standard used to describe the size of a wire. The large the AWG number, the smaller (thinner) the described wire.

Ampacity: The amount of current (measured in amperes) that a conductor can carry without overheating.

Ampere (or amp): Unit of current measurement. The amount of current that will flow through a one ohm resistor when one volt is applied.

Ampere-hour: The quantity of electricity equal to the flow of a current of one ampere for one hour.

Amplifier: A device that increases the voltage output (strength) of all signals received on a line. Amplifiers work on analog lines. Contrast with repeater.

Amplitude modulation: A transmission technique in which the amplitude of the carrier is varied in accordance with the signal.

GLOSSARY *(cont.)*

Amplitude: The size, in voltage, of signals in a data transmission. This voltage level is known as the amplitude.

Amplitude distortion: An unwanted change in the signal voltage which alters the data transmitted.

Analog: A format that uses continuous physical variables such as voltage amplitude or frequency variation to represent information. Contrast with digital.

Angle of incidence: The angle that a light ray striking a surface makes with a line perpendicular to the reflecting surface.

Angular misalignment: The loss of optical power caused by deviation from optimum alignment of fiber to fiber or fiber to waveguide.

Annunciator: A sound generating device that intercepts and speaks the condition of circuits or circuits' operations. A signaling device that gives a visual or audible signal (or both) when energized.

Anode: The positive electrode in an electrochemical cell (battery) toward which current flows.

ANSI: See: American National Standards Institute.

ANSI Character Set: The characters that include the 128 character ASCII character set and the 128 character Extended ASCII character set.

ANSI-568: Commercial Building Telecommunications Wiring Standard. See EIA-568.

Append: To change a file or program by adding to its end.

Application Layer: The top-most layer (Layer-7) in the OSI Reference Model providing such communication services as electronic mail and file transfer. It is generally defined as the top layer of a network protocol stack.

Application: A program that performs a user function. Synonymous with program.

Approved: Acceptable to the authority that has jurisdiction.

Approved Ground: A grounding bus or strap in a building that is suitable for connecting to data communication equipment. It includes a grounding subsystem, the building's electrical service conduit and a grounding conductor. See also EIA 607 and the National Electrical Code.

Arbitration: A set of rules for determining handling and priority of multiple communication sessions. Arbitration is less desirable than negotiation.

Architecture (computer): The conceptual design of computer hardware.

Architecture: The relationship of the physical parts of a computer or network that is typically labeled by that relationship. For example, the Motorola 68000 architecture.

GLOSSARY *(cont.)*

Archive: A procedure for transferring data from a storage disk, diskette or memory area to an external or removable storage medium.

Armoring: An additional protection layer on a cable to provide increased protection for abnormal environments. Often made of plastic coated steel it may also be corrugated for flexibility.

Array: A collection of photovoltaic (PV) modules, electrically wired together and mechanically installed in their working environment.

Artificial Intelligence (AI): A computer's ability to perform functions usually associated with human thought processes. For example, reasoning, learning, and correction or self-improvement. The current useable AI is Expert Systems.

ASCII: See American Standard Code for Information Interchange.

ASIC: Application Specific Integrated Circuit. A chip that is custom designed for a particular application.

Assembler: A language that uses symbolic machine language statements that have a one-to-one correlation to computer instructions.

Async: Asynchronous. A data transmission method that sends one character at a time. Contrasted with the synchronous methods, which send a packet of data and then resynchronize their clocks. Asynchronous also refers to commands, such as in windowing environment, that may be sent without waiting for a response from the previous command.

Asynchronous Transfer Mode (ATM): A method for the dynamic allocation of bandwidth using a fixed-size packet (called a cell). ATM is also known as "fast packet" and is an emerging WAN and LAN standard.

AT Command Set: The de facto modem command set developed by Hayes Microcomputer Products, Inc.

ATM: Automated Teller Machine or see Asynchronous Transfer Mode.

Attentuation coefficient: Characteristic of the attenuation of an optical fiber per unit length, in dB/km.

Attenuation: A general term used to denote the loss in strength of power between that transmitted and that received. This loss occurs through equipment, lines, or other transmission devices. It is usually expressed as a ratio in dB (decibel).

Attenuation Characteristic: As a signal travels on a cable, it gets weaker or attenuates. The attenuation characteristic of the medium is the rate at which it gets weaker.

Attenuator: A device that reduces signal power in a fiber optic link by inducing loss.

Authenticate (verb): The function of verifying the identity of a person or process.

Authentication: (1) The verification of the identity of a person or process. (2) the code used to identify a person or process.

GLOSSARY *(cont.)*

Authorization: Determining if a person or process is able to perform a particular action. Contrast with authentication.

Authorization Code: A multiple digit number or alphanumeric string entered to identify the user's authorization and/or level of authority for use of a system.

Autoexecute: The ability of an operating system to run certain programs without user intervention.

Automatic: Self-acting. Operating by its own mechanism, based on a non-personal stimulus.

Automatic Restart: A function in which a process can automatically restart from a point of failure. The halt may have been the result of a power or circuit failure or an interrupt generated by a user.

Average Power: The average over time of a modulated signal.

AWG: See American Wire Gauge.

Azimuth: Horizontal angle measured from true north.

B: Byte.

b: bit.

Babble: Crosstalk from multiple channels or circuits.

Back reflection, optical return loss: Light reflected from the cleaved or polished end of a fiber caused by the difference of refractive indices of air and glass. Typically 4 percent of the incident light. Expressed in dB relative to incident power.

Back-up (noun): The copy of data that results from a back-up procedure. Sometimes refers to the storage media that contains that data.

Back-up (verb): To copy a file, directory, or column onto another storage device so that the data is retrievable if the original source is accidentally corrupted or destroyed.

Backboard: A wooden (or metal) panel used for mounting equipment usually on a wall.

Backbone: The main connectivity device of a distributed system. All systems that have connectivity to the backbone will connect to each other. This does not stop systems from setting up private arrangements with each other to bypass the backbone for cost, performance or security.

Backbone Cable: A main Cable run vertically (or horizontally) in a building to provide wire connectivity to separate areas in the building. It is not designed for direct system access.

Backbone Closet: Space provided in a building for terminating pairs of wire and connecting the backbone cable to systems in that area. See also**:** Telecommunications Closet.

Backbone Wiring: See backbone cable.

GLOSSARY *(cont.)*

Backfeed Pull: A cable pulling method that starts in the middle of a conduit and feeds the cable, in one direction and then the other, from that location.

Background Noise: Extra signals that are found on a circuit, line or channel.

Backscattering: The return of a portion of scattered light to the input end of a fiber; the scattering of light in the direction opposite to its original propagation.

Backup Link: An alternate circuit that is not used until, or if, the primary link fails.

Ballast: An electrical circuit component used with fluorescent lamps to provide the voltage necessary to strike the mercury arc within the lamp, and then to limit the amount of current that flows through the lamp.

Balun: Balanced/unbalanced. Refers to an impedance-matching device used to connect balanced twisted-pair cabling with unbalanced coaxial cable, best known in the IBM cabling system.

Band Pass Filter: An electronic device that filters out all signals except those one or more selected frequency ranges.

Bandwidth: Technically, the difference, in Hertz (Hz), between the highest and lowest frequencies of a transmission channel. Usually identifies the capacity or amount of data that can be sent through a given circuit.

BASIC: Beginner's All-purpose Instruction Code. A computer compiler language; similar to FORTRAN.

Basic Input/Output System (BIOS): A set of programs, usually in firmware, that lets each computer's central processing unit communicate with printers, disks, keyboards, consoles, and other attached input and output devices.

Battery: A device that converts chemical energy into electrical current.

Battery cycle life: The number of cycles that a battery can undergo before failing.

Battery self-discharge: Loss of chemical energy in a battery that is not under load.

Baud: A unit of signaling speed. The speed in Baud is the number of discrete conditions or signal elements per second. If each signal event represents only on bit condition, then Baud is the same as bits per second. Baud rarely equals bits per second.

Baud Rate: The rate at which data is transferred over an asynchronous RS-232 serial connection.

BBS: See: Bulletin Board System.

Beamsplitter: An optical device, such as a partially reflecting mirror, that splits a beam of light into two or more beams and that can be used in fiber optics for directional couplers.

Bearing Wall: A wall supporting a load other than its own weight such as the next floor above. Also called a load bearing wall.

GLOSSARY *(cont.)*

Bend loss: A form of increased attenuation in a fiber that results from bending a fiber around a restrictive curvature (a macrobend) or from minute distortions in the fiber (microbends).

Bend Radius: the radius a cable can bend without risking breakage or increasing attenuation.

BER: Basic Encoding Rules. Standard rules for encoding data units described in ASN.1. Sometimes incorrectly lumped under the term ASN.1, which properly refers only to the abstract syntax description language, not the encoding technique.

Beta Test: The stage at which a new product is tested under actual conditions. These tests are usually performed by selected product users. See also Alpha Test.

Binary: The base-2 number system using only the symbols 0 and 1. since 0 and 1 can be represented as on and off, or negative and positive charges, most computers do their calculations in binary.

BIOS: See Basic Input/Output System.

Bit: A binary digit; must be either a 0 or a 1. It is the smallest unit of information and indicates one of the two electrical states: off (O) or on (1) in a computer.

Bit Error Rate (ber): In testing, the ratio between the total number of bits transmitted in a given message and the number of bits in that message received in error. A measure of the quality of a data transmission, usually expressed as a number referred to a power of 10; e.g., 1 in 10 over 5.

Bit Rate: See bits per second.

Bit-mapped: Refers to a display screen on which a character or image is generated and refreshed according to a binary matrix (bit map) at a specific location in memory.

Bits Per Second (bps): Basic unit of measurement for serial data transmission capacity, abbreviated as k bps, or kilobit/s, for thousands of bits per second; m bps, or megabit/s, for millions of bits per second; g bits, or gigabit/s for billions of bits per second; t bps, or terabit/s, trillions of bits per second.

Black Box: Any electronic device that is not understood. Also used to describe a device who's detailed components or functions can be ignored in a discussion.

Block: A unit of stored or transmitted data. It is usually in a standard size such as 512, 1024 or 4,096 bytes of data.

Blocking diode: A diode used to prevent current flow in a photovoltaic array during times when the array is not producing electricity.

Bonding: A very-low impedance path accomplished by permanently joining non-current-carrying metal parts. It is done to provide electrical continuity and to offer the capacity to safely conduct any current.

Bonding jumper: A conductor used to assure the required electrical connection between metal parts of an electrical system.

GLOSSARY *(cont.)*

Bonding Conductor: The conductor that connects the noncurrent- carrying parts of electrical equipment, cable raceways, or other enclosures to the approved system ground conductor.

Boot, or bootstrap (computer): A short series of instruction codes that program a computer to read other codes. A machine that has no program at all in its memory cannot even read data.

Boot up: See boot.

Braid: A woven group of filaments that covers one or more insulated conductors, particularly in coax cabling.

Branch circuit: Conductors between the last overcurrent device and the outlets.

Branch circuit, general purpose: A branch circuit that supplies outlets for lighting and power.

Branch circuit, individual: A branch circuit that supplies only one piece of equipment.

Branch circuit, multiwire: A branch circuit having two or more ungrounded circuit conductors, each having a voltage difference between them, and a grounded circuit conductor (neutral) having an equal voltage difference between it and each ungrounded conductor.

Breakout Box: A device that allows access to individual points on a physical interface connector for testing and monitoring. it is often used to troubleshoot RS-232 circuits and cables.

Breakout Cable: A multifiber cable where each fiber is protected by an additional jacket and strength element beyond that provide for the overall cable.

Broadband: A transmission medium capable of supporting a wider range of frequencies than that required for a single communication channel. It can simultaneously carry multiple signals by dividing the total capacity of the medium into multiple, independent bandwidth channels, where each channel operates only on specific range of frequencies. Broadband implies the use of a frequency agile modem to select the correct channel rather than direct modulation as used in baseband. See also: baseband.

Buffer: In systems, a portion of memory designated for temporary storage of data. In cabling, a protective material on the fiber coating to protect the fiber-optic cable. See also tight buffer or loose buffer.

Building: A structure which is either standing alone, or cut off from other structures by fire walls.

Building Core: That portion of any building devoted to elevators, stairwells, vertical plumbing, rest rooms, vertical electrical and communications cables and equipment.

Bulletin Board System (BBS): A computer, and related software, which typically provides electronic messaging services, archives of files, and any other services or activities of interest to the bulletin board systems's operator.

GLOSSARY *(cont.)*

Bus Bar: The heavy copper or aluminum bar used to carry currents in switchboards.

Bus (data): The primary signal route, inside a computer, which can have several devices connected, letting them transmit and receive data at the same time. For example, IBM's Micro Channel Architecture.

Bus (network): The main (multiple access) network cable or line that connects network stations. Also refers to a network topology of multiple stations communicating directly with the same cable with terminators at both ends, like an Ethernet or token bus.

Busy Tone: A single tone that is interrupted at 60 ipm (impulses per minute) to indicate that the terminal point of a call is already in use.

Bypass Diode: A diode connected in parallel with a block of parallel modules to provide an alternate current path in case of module shading or failure.

Byte: One character of information, usually 8-bits.

Cable plant, fiber optic: The combination of fiber optic cable sections, connectors, and splices forming the optical path between two terminal devices.

Cable Assembly: A completed cable, with its connectors and hardware, that is ready for installation.

Cable Patch Panel: A passive device frequently located in the intermediate distribution facility (IDF) or satellite equipment room to offer easy circuit cross connections. It is used to connect two sets of wire (e.g., the wire from the IDF to the office and the wire between the IDFs).

Cable System: All the cables and devices used to interconnect stations; often called the premises network.

Cache: A portion of a computer's RAM reserved to act as a temporary memory for items read from a disk. They become instantly available to the user.

CAD/CAM: Computer aided design/computer aided manufacture. Software/hardware combinations for the automation of engineering environments.

Campus Backbone: The primary wiring that travels through a campus. Buildings and their internal networks connect to the backbone for communications with each other.

Capacitor: An electrical device which causes the current in a circuit to lead the voltage, the opposite effect of induction.

Carrier Frequency: Frequency of the carrier wave that is modulated to transmit signals.

Category Cable: Cable that complies with EIA/TIA TSB 36 and is rated category 1 thru 5. The higher the category number, the better the cable will be at carrying high speed signals. Category wire may be shielded, but is always 100 ohm.

Category Devices: Devices, such as patch panels and wall jacks, that comply with EIA/TIA TSB40. See category cable.

GLOSSARY *(cont.)*

Cathode: The negative electrode in an electrochemical cell.

CATV: An abbreviation for community antenna television or cable TV.

Central Office: The site where communications common carriers (telephone companies) terminate customer lines and house the equipment that interconnects these lines.

Character Set: A collection of characters, such as ASCII or EBCDIC, used to represent data in a system. These character are typically available on a keyboard or through a printer.

Charge controller: A device that controls the charging rate and state of charge for batteries. See Charge Rate.

Charge Rate: The rate at which a battery is recharged. Expressed as a ratio of battery capacity to charge current flow, for instance, C/5.

Charge controller: A device that controls the charging rate and state of charge for batteries. See Charge Rate.

Charge Rate: The rate at which a battery is recharged. Expressed as a ratio of battery capacity to charge current flow, for instance, C/5.

Chromatic dispersion: The temporal spreading of a pulse in an optical waveguide caused by the wavelength dependence of the velocities of light.

Circuit breaker: A device used to open and close a circuit by automatic means when a predetermined level of current flows through it.

Circuit Switching: A communications method in which a dedicated path is identified by switching a signal to the wires that will connect the two hosts. The telephone system is an example of a circuit switched network. See also: connection-oriented, connectionless, packet switching.

Cladding: The outer concentric layer that surrounds the fiber core and has a lower index of refraction.

Clock: The timing signals used in data communications or the source of those timing signals.

CMOS: Complimentary Metal Oxide Semiconductor.

CO: See Central Office. Location where communications common carriers terminate customer lines and house the equipment that interconnects these lines.

Co-Processor: An additional central logic unit which performs specific tasks while the main unit executes its primary tasks. Frequently, these chips are added to speed up mathematical tasks or perform I/O functions.

Coaxial Cable: A transmission medium noted for its wide bandwidth and for its low susceptibility to interference. It is made up of an outer woven conductor which surrounds the inner conductor. The conductors are commonly separated by a solid insulating material.

COBOL: Common business-oriented language. One of the first standardized computing languages.

GLOSSARY *(cont.)*

Code: (verb) To write compilable software. (noun) The compilable software that was written. Also the rules specifying how data may be represented in a particular system. Slang for program.

Common mode: Placed upon both sides of an amplifier at the same time.

Common Carrier: An organization in the business of providing regulated telephone, telegraph, telex, and data communications services.

Communications Protocol: The rules used to control the orderly exchange of information between stations on a data link or on a data network or system. Also called line discipline or protocol.

Communications Satellite: An earth satellite designed to act as a communications radio relay and usually positioned in geosynchronous orbit 35,800 kilometers (23,000 miles) above the equator so that it seems to be stationary in space.

Component: A type of network element like routers, computers, operating systems, gateways, etc.

Compression: A method to reduce the number of bits required to represent data.

Computer Network: An interconnection of computer systems, terminals, communications facilities, and data collecting devices.

Concealed: Made inaccessible by the structure or finish of the building.

Concentrator: 1) A photovoltaic module that uses optical elements to increase the amount of sunlight incident on a PV cell. 2) A communications device that offers the ability to concentrate many lower-speed channels into and out of one or more high-speed channels.

Conductor: A substance which offers little resistance to the flow of electrical currents. Insulated copper wire is the most common form of conductor.

Conduit body: The part of a conduit system, at the junction of two or more sections of the system, that allows access through a removable cover. Most commonly known as condulets, LBs, LLs, and LRs.

Continuous load: A load whose maximum current continues for three hours or more.

Conduit body: The part of a conduit system, at the junction of two or more sections of the system, that allows access through a removable cover. Most commonly known as conduits, LBs, LLs, LRs, etc.

Conferencing: A term used for communication software that allows participants to post notes. It is unlike electronic mail since participants do not have to be explicitly addressed. A primary function of a bulletin board. It also lets multiple users simultaneously interact online as a computer conference call.

Configuration: Settings that control the way a system or service will operate. Also the combined services and/or equipment that make up a communications system.

GLOSSARY *(cont.)*

Connect Time: A measure of system usage. It is the interval during which the user was on-line for a session.

Continuous load: A load whose maximum current continues for three hours or more.

Controller: A device or group of devices that control (in a predetermined way) power to a piece of equipment.

Conversion Efficiency: The ratio of the electrical energy produced by a photovoltaic cell to the solar energy received by the cell.

Copper Distributed Data Interface (CDDI): A variation of FDDI that uses Category 5 unshielded twisted pair copper wire. See also FDDI, TP-PMD.

Core: The central, light-carrying part of an optical fiber; it has an index of refraction higher than that of the surrounding cladding.

Core storage: Binary memory storage, made up of tiny magnetic elements.

CPU: Central Processing Unit. See also microprocessor.

Cross Connection: The wire connections running between wiring terminal blocks. Cross connections let pairs in one cable connect to pairs in another cable. See also patch panel, on the two sides of a distribution frame, or between binding posts in a terminal.

Cross-sectional area: The area (in square inches or circular mils) that would be exposed by cutting a cross-section of the material.

Crosstalk: The unwanted energy transferred from one circuit or wire to another which interferes with the desired signal. Usually caused by excessive inductance in a circuit.

CRT: Cathode Ray Tube (also generic reference to a terminal).

Crystalline Silicon: A type of PV cell made from a single crystal or polycrystalline slice of silicon.

Current: The flow of electricity in a circuit, measured in amperes.

Cursor: A movable underline, rectangular-shaped block of light, or an alternating block of reversed video on the screen of a display device, usually indicting where the next character is to be entered.

Customer Access Line Charge (CALC): The FCC-imposed monthly surcharge added to all local lines to recover a portion of the cost of telephone poles, wires, etc., from end users. Before deregulation, a large part of these costs were financed by long distance users in the form of higher charges.

Customer Premise Equipment (CPE): Equipment, usually including wiring located within the customer's part of a building.

Cut-out box: A surface mounted electrical enclosure with a hinged door.

GLOSSARY *(cont.)*

Cutback method: A technique for measuring the loss of bare fiber by measuring the optical power transmitted through a long length then cutting back to the source and measuring the initial coupled power.

Cutoff wavelength: The wavelength beyond which single-mode fiber only supports one mode of propagation.

Cutoff Voltage: The voltage at which the charge controller disconnects the array from the battery. See Charge Controller.

Cyberspace: A term to describe the "world" of computers, and the society that gathers around them.

Daisy chaining: The connection of multiple devices in a serial fashion. Daisy chaining can save on transmission facilities. If a device malfunctions all of the devices daisy chained behind it are disabled.

Data rate: The number of bits of information in a transmission system, expressed in bits per second (bps), and which may or may not be equal to the signal or baud rate.

Data Base: A large, ordered collection of information.

Data Communications: The interchange of data messages from various sources are accumulated.

Data Communications Equipment: The equipment which provides the functions of interfacing between data terminal equipment and a communications channel DCE is normally a modem.

Data Link: A transmission path directly connecting two or more stations (a station may be a terminal, terminal controller, front end processor, or other type of digital equipment).

Data Terminal: A station in a system capable of sending and/or receiving data signals.

Data Transmission: The sending of data from one place for reception elsewhere.

dB: Decibel referenced to a microwatt.

dBm: Decibel referenced to a milliwatt.

Decibel: 1) A standard logarithmic unit for the ratio of two powers, voltages, or currents. In fiber optics, the ratio is power. 2) Unit for measuring relative strength of a signal parameter such as power or voltage.

$$dB = 10 \log^{10} \left(\frac{P^1}{P^2}\right)$$

Decimal: a digital system that has ten states, 0 through 9.

Deep Cycle: Battery type that can be discharged to a large fraction of capacity. See Depth of Discharge.

GLOSSARY *(cont.)*

Default Route: A routing table entry which is used to direct any data addressed to any network numbers not otherwise listed in the routing table.

Degauss: To remove residual permanent magnetism.

Depth of Discharge (DOD): The percent of the rated battery capacity that has been withdrawn.

Detector: An optoelectronic transducer used in fiber optics for converting optical power to electric current. In fiber optics, usually a photodiode.

Device (Also used as wiring device): The part of an electrical system that is designed to carry, but not use, electrical energy.

Diagnostics: Programs or procedures used to test a piece of equipment, a communications link or network.

Dial Tone: A tone indicating that automatic switching equipment is ready to receive dial signals.

Diameter-mismatch loss: The loss of power at a joint that occurs when the transmitting half has a diameter greater than the diameter of the receiving half. The loss occurs when coupling light from a source to fiber, from fiber to fiber, or from fiber to detector.

Diffuse Radiation: Radiation received from the sun after reflection and scattering by the atmosphere.

Digital: Communications procedures, techniques, and equipment whereby information is encoded as either binary "1" or "0". Also the representation of information in discrete binary form.

Diode: Electronic component that allows current flow in one direction only.

Dip Switch: A dual in-line package switch. It has two parallel rows of contacts that let the user switch electrical current through a pair of those contacts to on or off. They are used to reconfigure computer components and peripherals. See also jumpers.

Direct current (DC): Electrical current which flows in one direction only.

Disconnecting means: A device which disconnects a group of conductors from their source of supply.

Discrete Access: An access method used in star LANs: each station has separate (discrete) connections it uses for the LAN's switching capability. Contrast with shared access.

Disk: A rotating disk covered with magnetic material, used for storage of data.

Disk Operating System: A program or set of programs that tells a disk-based computer system to schedule and supervise work, manage computer resources, and operate and control its peripheral devices.

Disk Server: A LAN device that lets multiple users access sections of its disks for creating and storing files. Contrast with a file server, which allows users to share files.

GLOSSARY *(cont.)*

Dispersion: A general term for those phenomena that cause a broadening or spreading of light as it propagates through an optical fiber. The three types are modal, material, and waveguide.

Distortion: An unwanted change to signal caused by outside interference or by imperfections of the transmission system. Often caused by excess capacitance.

Distributed Data Processing: The processing of information in separate locations as on a local area network. This is a more efficient use of processing power since each CPU can do a certain task.

Distributed Database: A collection of several different data repositories that looks like a single database to the user.

Distribution Block or Frame: Centralized equipment where wiring is terminated and cross-connections are made. See also Intermediate Distribution Frame and Main Distribution Frame.

Domain: A zone or part of a naming hierarchy. An Internet domain name consists of a sequence of names separated by periods, as in "roadie.cs.arg3.com."

DOS: See Disk Operating System.

Down Time: The total time a system is out of service due to equipment failure.

Download: To make a copy of a file from a central service (or server) onto a local computer. Contrast with a load and execute operation on a LAN.

DRAM: Dynamic Random Access Memory. Contrast with Static RAM.

Driver: The amplifier stage preceding the output stage.

Drop (noun): A portion of cable that connects a user station to a network. Also refers to the jack that is the point of contact for the cable drop. The user sees a drop as a network connection.

Drop (verb): To logically disconnect part or all of a signal whether intentionally or unintentionally.

Drop Cable: Cable that provides access to and from a network system. Possibly the cable from a transceiver or an individual line in a multi-drop situation. Also the cable from a wall-mounted faceplate or jack to a user's system.

Dumb Terminal: A term used to describe an asynchronous, ASCII terminal that, although it may be "intelligent" in many of the functions it provides, uses no communication protocol. It may operate at speeds up to 19.2 Kbps.

Duplex cable: A two-fiber cable suitable for duplex transmission.

Duplex transmission: Transmission in both directions, either one direction at a time (half duplex) or both directions simultaneously (full duplex).

GLOSSARY *(cont.)*

Duplex: Simultaneous two-way independent transmission.

Earth Station: Ground-based equipment used to communicate via satellites.

Edge-emitting diode (E-LED): A LED that emits from the edge of the semiconductor chip, producing higher power and narrower spectral width.

EDP: Electronic data processing.

EEPROM: See Electrically Erasable Programmable Read-Only Memory.

EIA: See Electronics Industries Association.

EIA/TIA 568: The commercial building wiring standard. It defines a generic wiring system for a multi-product, multi-vendor environment. Widely considered the most important standard for building wiring.

EIA/TIA 570: The residential and light commercial building wiring standard. It is the wiring standard for single and multi-family residential and mixed use facilities.

EIA-232: Standard interface definition for serial devices. better known as RS-232. RS-232-E is the current version of the standard.

Electrically Erasable Programmable Read-Only Memory (EEPROM): A memory that can be electronically programmed and erased, but which does not need a power source to hold the data.

Electrolyte: A liquid or paste in which the conduction of electricity is by a flow of ions.

Electromagnetic Interference (EMI): The energy given off by electronic circuits and picked up by other circuits; based on the type of device and operating frequency. EMI effects can be reduced by shielding and other cable designs. Minimum acceptable levels are detailed by the FCC. See also Radio Frequency Interference.

Electronic Data Interchange (EDI): A standard system of exchanging order and billing information between computers in different companies.

Electronic Mail (email): A system that lets computer users exchange messages with other computer users (or groups of users). The messages go to an email server instead of directly to the end recipient.

Electronic Switch: A modern programmable switch (also called ESS, for Electronic Switching System). It contains only solid state electronics, unlike older mechanical switches.

Electronics Industries Association (EIA): A US trade organization that issues its own standards and contributes to ANSI. Best known for its development RS-232 and the building wiring standard, 568. Membership includes US manufacturers.

Email: See Electronic Mail.

Email Address: The address that is used to send electronic mail to a specific destination. For example: jdoe@isp.com.

GLOSSARY *(cont.)*

EMI: See Electromagnetic Interference.

EMP: Electromagnetic Pulse.

Enclosed: Surrounded by a case, housing, fence, or walls that prevent unauthorized people from contacting the equipment.

Encryption: Manipulation of a packet's data in order to prevent any but the intended recipient from reading that data.

End User: The human source and/or destination of information sent through the communications system.

EOM: End of Message.

EPROM: Erasable Programmable Read Only Memory.

Equalization: The process of restoring all cells in a battery to an equal state of charge.

ESCON: IBM standard for connecting peripherals to a computer over fiber optics. Acronym for Enterprise System Connection.

Ethernet: A 10-Mbps, coaxial standard for LANs, initially developed by Xerox and later refined by Digital, Intel and Xerox (DIX). All nodes connect to the cable where they contend for access via CSMA/CD. Also slang for the coaxial cable that carries the standard.

Ethernet Controller: A device that controls a computer's access to Ethernet services. The CSMA/CD protocols are used by the controller to free the CPU.

Ethernet Meltdown: The result of an event that causes saturation, or near saturation, on an Ethernet. It usually comes from illegal or misrouted packets and normally runs for a short time only.

Event: The occurrence of a particular change in the state of an managed object.

Event Log: A record of significant events.

Excess loss: In a fiber-optic coupler, the optical loss from that portion of light that does not emerge from the nominally operational ports of the device.

Exchange: The collection of equipment in a communications system that controls the connection of incoming and outgoing lines. Also known as central office.

EXEC: Executable.

Exposed: Able to be inadvertently touched or approached.

Extrinsic loss: In a fiber interconnection, that portion of loss that is not intrinsic to the fiber but is related to imperfect joining, which may be caused by the connector or splice.

Facsimile: A copy. Also the transmission of documents via communications circuits. This is done by using a device which scans the original document, transforms the image into coded signals and reproduces documents it receives. Also called fax.

GLOSSARY *(cont.)*

Fading: A situation, generally of microwave or radio transmission, where external influences cause a signal to be deflected or diverted away from the target receiver. Also the reduction in intensity of the power of a received signal.

Fall time: The time required for the trailing edge of a pulse to fall from 90% to 10% of its amplitude; the time required for a component to produce such a result. "Turnoff time". Sometimes measured between the 80% and 20% points.

FAQ: Frequently Asked Questions.

Farad: The unit of measurement of capacitance.

Fault: A condition that causes any physical component of a system to fail to perform in acceptable fashion.

Fault Tolerance: The ability of a program or system to operate properly even if a failure occurs.

FDDI: See Fiber Distributed Data Interface.

FDDI II: See Fiber Distributed Data Interface.

Federal Communications Commission (FCC): The government agency established by the Communications Act of 1934 which regulates interstate communications.

Feeder: Circuit conductors between the service and the final branch circuit over current device.

Ferrule: A precision tube that holds a fiber for alignment for interconnection or termination. A ferrule may be part of a connector or mechanical splice.

Fiber identifier: A device that clamps onto a fiber and couples light from the fiber by bending, to identify the fiber and detect high-speed traffic of an operating link or a 2 kHz tone injected by a test source.

Fiber tracer: An instrument that couples visible light into the fiber to allow visual checking of continuity and tracing for correct connections.

Fiber Channel: A switched datalink technology released as a draft by ANSI. It uses coaxial cable and shielded twisted pair at speeds of 133 Mbps to 1 Gbps.

Fiber Distributed Data Interface (FDDI): A high-speed (100Mb/s) LAN standard. The underlying medium is fiber optics, and the topology is a dual-attached, counter-rotating token rings,. FDDI II is a draft standard for revising FDDI to carry multi-media signals at 100 Mbps. See also: Local Area Network, token ring.

Fiber Optics: A technology that uses light as a digital information carrier. Fiber optic cables are direct replacement for conventional cables and wire pairs. They occupy far less physical space and are immune to electrical interference.

File (computer): A block of data.

GLOSSARY *(cont.)*

File Server: In local networks, a station dedicated to providing file and mass data storage services to the other network stations.

File Transfer: The copying of a file from one computer to another over a network or dial-up circuit.

File Transfer Protocol (FTP): A TCP/IP protocol that lets a user on one computer access, and transfer files to and from, another computer over a network. FTP is usually the name of the program the user invokes to accomplish this task.

Filter: A combination of circuit elements which is specifically designed to pass certain frequencies and resist all others.

Firmware: Permanent or semi-permanent micro-instruction control for a user-oriented function.

Flame: A strong opinion and/or criticism of something, usually a frank inflammatory statement, in an electronic mail message. It is common to precede a flame with an sign of pending fire (as in "FLAME ON!").

Float Charge: The charge to a battery having a current equal to or slightly greater than the self discharge rate.

Floating Point: A native data type on most operating systems that can have numbers after the decimal point. Contrast with an integer.

Fluorescence: The emission of light by a substance when exposed to radiation or the impact of particles. The effect ceases in a fraction of a second once the source of radiation or particles is removed.

FM: Frequency modulation.

FO: common abbreviation for fiber optic.

Footprint: (1) The space a device occupies on a desk or work surface. (2) The precise area of the earth in which a satellite communications signal can be received.

Four Wire Circuits: Telephone circuits which use two separate one-way transmission paths of two wires each, as opposed to regular local lines which usually only have two wires to carry conversations in both directions.

Four-Wire Channel: A circuit containing two pairs of wire (or their logical equivalent) for simultaneous (i.e., full-duplex) two-way transmission.

Frame Relay: A faster form of packet switching that is accomplished with smaller packet sizes and less error checking.

Frequency modulation: A method of transmission in which the carrier frequency varies in accordance with the signal.

Frequency: The number of times per second a signal regenerates itself at a peak amplitude. It can be expressed in hertz (Hz), kilohertz (Hz), megahertz (MHz), etc.

GLOSSARY *(cont.)*

Frequency Division Multiplexing (FDM): A method of dividing an available frequency range into subparts, each having enough bandwidth to carry one channel.

Fresnel reflection: The reflection that occurs at the planar junction of two materials having different refractive indices; Fresnel reflection is not a function of the angle of incidence.

Fresnel reflection loss: Loss of optical power due to Fresnel reflections.

Full Duplex: A circuit that lets messages flow in both direction at the same time. Contrast with half-duplex where only one side can transmit at a time.

Fusion splicer: An instrument that splices fibers by fusing or welding them, typically by electrical arc.

GAN: Global Area Network.

Gap loss: Loss resulting from the end separation of two axially aligned fibers.

Gassing: Gas by-products produced when charging a battery. Also, termed out-gassing.

Gateway: The original Internet term for a router or more accurately, an IP router. In current usage, "gateway" and "application gateway" refer to translating systems that convert data traveling from one environment to another.

GB: Gigabit (or gigabyte) backbone. The gigabit backbone is an effort to increase the speed of the Internet to one Gbps.

Gb: Gigabit. One billion bits of information.

Gbyte: Gigabyte. One billion bytes of data.

GFLOPS: Billion Floating Operations Per Second.

Giga: A prefix that means one billion.

Graded-index fiber: An optical fiber whose core has a nonuniform index of refraction. The core is composed of concentric rings of glass whose refractive indices decrease from the center axis. The purpose is to reduce modal dispersion and thereby increase fiber bandwidth.

Grid: Term used to describe an electrical utility distribution network.

Ground: An electrical connect (on purpose or accidental) between an item of equipment and the earth.

Hacker: A person who delights in having an intimate understanding of the internal workings of computers and computer networks.

Half Duplex: A circuit for transmitting or receiving signals in one direction at a time.

Hand Shaking: The exchange of predetermined control signals for establishing a session between data sets.

Hard copy (computer): A printout on paper or cards.

GLOSSARY *(cont.)*

Hardware (computer): The physical computer and related machines.

Hardwire: To wire or cable directly between units of equipment.

Harmonic: 1) A sinusoid which has a frequency which is an integral multiple of a certain frequency. 2) The full multiple of a base frequency.

HDTV: High Definition Television.

Hertz (Hz): International standard unit of frequency. Replaces the identical older "Cycles-per-second."

Hexadecimal: A number system with 16 members represented by 0 through 9 followed by A through F. Each character identifies four bit or a half-byte. Also called hex.

Host: (1) A computer that provides services directly to users, i.e. the user's computer. In TCP/IP, an IP addressed device. (2) A large computer that serve many users, i.e. a minicomputer or mainframe.

Host Address: The part of an Internet address that designates which node on the (sub)network is being addressed. Also called host number.

Host Computer: See host.

Hub: A device which connects to several other devices usually in a star topology. Also called: concentrator, multiport repeater or multi-station access unit (MAU).

Hybrid: An electronic circuit that uses different cable types to complete the circuit between systems.

Hz: See Hertz.

I/O or Input-Output: Related to the process of getting data into and out of a computer or processor.

Identified (for use): Recognized as suitable for a certain purpose, usually by an independent agency, such as U. L.

IDP: Integrated detector/preamplifier.

IEEE: Institute of Electrical and Electronic Engineers (US).

Impedance: The effects placed upon an alternating current circuit by induction, capacitance, and resistance. Total resistance in an AC circuit.

Index of refraction: The ratio of the velocity of light in free space to the velocity of light in a given material. Symbolized by n.

Index-matching material: A material, used at optical interconnection, having a refractive index close to that of the fiber core and used to reduce Fresnel reflections.

Inductance: The characteristic of a circuit that determines how much voltage will be induced into it by a change in current of another circuit.

GLOSSARY *(cont.)*

Insertion loss: The loss of power that results from inserting a component, such as a connector or splice, into a previously continuous path.

Integrated circuit or IC: A circuit in which devices such as transistors, capacitors, and resistors are made from a single piece of material and connected to form a circuit.

Interconnect: (1) The arrangement that allows the connection of customer's communications equipment to a common carrier network. (2) The generic term for a circuit administration point that allows routing and rerouting of signal traffic.

Interface: The point that two systems, with different characteristics, connect.

Internet: (Note the capital "I") The largest Internet in the world including large national backbone nets (such as MILNET, NSFNET, and CREN) and many regional and local networks world- wide. The Internet uses the TCP/IP suite. Networks with only email connectivity are not considered on the Internet.

Interoperability: The ability of software and hardware on multiple machines from multiple vendors to communicate meaningfully.

Intrastate: Any connection made that remains within the boundaries of a single state.

Inventor: A device for changing direct current into alternating current by alternately switching DC in inverted polarity.

Ion: An atom or molecule that has acquired a charge by gaining or losing one or more electrons.

IRQ: Interrupt Request

Isolated: Not accessible unless special means of access are used.

Isolation transformer: A one to one transformer that is used to isolate the equipment at the secondary from earth ground.

Jack: A receptacle (female) used with a plug (male) to make a connection to in-wall communications cabling or to a patch panel.

Jacket: The protective and insulating outer housing on a cable. Also called a sheath.

Jumper: Patch cable or wire used to establish a circuit, often temporarily, for testing or diagnostics. Also the devices, shorting blocks, used to connect adjacent exposed pins on a printed circuit board that control the functionality of the card.

Junction Box: A box, usually metal, that encloses cable connections for their protection.

K Band: A microwave band from 10.9 GHz to 36GHz.

KB: Kilobyte. One thousand bytes.

Kb: Kilobit. One thousand bits.

Kbps: Kilobits per second. Thousand bits per second.

GLOSSARY *(cont.)*

Kbyte: Kilobyte. One thousand bytes.

Kermit: A popular file transfer protocol developed by Columbia University. By running in most operating environments, it provides an easy method of file transfer. Kermit is NOT the same as FTP. See also File Transfer Protocol.

KU Band: The frequency band from 12 to 14 GHz that is used for satellite communications.

kW: The abbreviation for kilowatt, a unit of measurement of electrical power. One kilowatt is equal to one thousand watts.

L Band: Microwave and satellite communications frequencies in the 390 MHz to 1550 MHz range.

Lamp: A light source. Reference is to a light bulb, rather than a table lamp.

LAN: See Local area network.

LAN Adapter: An external device or card, for internal use that lets a device gain access to a local area network.

Laser: See Light Amplitude by Stimulated Emission of Radiation.

Laser Optical: The generic term for the system of recording data, sound or video on optical discs.

Launch cable: A known good fiber optic jumper cable attached to a source and calibrated for output power used for loss testing. This cable must be made of fiber and connectors of a matching type to the cables to be tested.

Layer: A modular portion of a stacked protocol that consists of one or more semi-independent protocols. Each layer builds on the layer beneath it and feeds information to the protocols in the layers above it. TCP/IP has five layers of protocols, and OSI has seven.

LED: Light emitting diode.

Level: An expression of the relative signal strength at a point in a communications circuit as compared to a standard.

Life-Cycle Cost: The estimated cost of owning and operating a system for the period of its useful life.

Light Amplitude by Stimulated Emission of Radiation (laser): A device that produces a very uniform, single frequency of light. Digital signals can be transmitted by turning it on and off rapidly. See also Fiber Optics.

Light Emitting Diode: A semiconductor diode that gives off light when current is passed through it.

Line: An electrical path between two points, usually a telco CO and the end user.

Line Driver: A short haul communications device for overcoming the RS-232 alleged distance limitation. See also short haul modem or limited distance modem.

GLOSSARY *(cont.)*

Link: The physical interconnection between two systems (sometimes called nodes) in a network. A link may consist of a data communications circuit or a direct cable connection.

Load: The amount of electric power used by any electrical unit or appliance at any given moment.

Local Area Network (LAN): A data network intended to serve an area of only a few square kilometers or less. Because the network is known to cover only a small area, optimizations can be made in the network signal protocols that permit higher data rates. See also Ethernet, Fiber Distributed Data Interface, token ring, Wide Area Network.

Local Loop: the local connection between the end user and the Class 5 central office.

Location, damp (damp location): Partially protected locations, such as under canopies, roofed open porches, etc. Also, interior locations that are subject only to moderate degrees of moisture, such as basements, barns, etc.

Location, dry (dry location): Areas that are not normally subject to water or dampness.

Location, wet (wet location): Locations underground, in concrete slabs, where saturation occurs, or outdoors.

Login: To gain access to a computer or network by identifying the acceptable user name and passing the required authentication procedure(s).

Long Haul: circuits spanning considerable distances.

Loss budget: The amount of power lost in a fiber optic link. Often used in terms of the maximum amount of loss that can be tolerated by a given link.

Loss, optical: The amount of optical power lost as light is transmitted through fiber, splices, couplers, and the like.

Machine language: Programs or data that is in a form that is immediately useable by the computer, usually Binary.

Macintosh: A computer made by Apple Computer that is characterized by the graphical, intuitive user interface.

Main Distribution Frame (MDF): The point where outside cables and backbone cables are cross connected. Usually the point at which all cables from intermediate distribution frames intersect.

Mainframe: A generic term of a large, multi-user, multi-tasking computer. Many use it to refer to IBM mainframe computers.

MAN: See Metropolitan Area Network

Mbps: Million bits per second.

Mbyte: Megabyte. Million bytes of information.

MDF: See Main Distribution Frame.

GLOSSARY *(cont.)*

Mechanical splice: A semipermanent connection between two fibers made with an alignment device and index matching fluid or adhesive.

Media: The plural of medium.

Media Interface Connector: The optical fiber connector which connects the fiber to the FDDI controller.

Medium: (1) Any substance that can be used for the propagation of signals, such as optical fiber, cable, wire, dielectric slab, water, air, or free space. (2) the material on which data is recorded; for example, magnetic type, diskette.

Metropolitan Area Network (MAN): A data network intended to serve an area approximating that of a large city. See also Local Area Network, Switched Multimegabit Data Service, Wide Area Network.

Microcomputer: A small-scale programmable machine that processes information. It generally has a single chip as its central processing unit and includes storage and input/output facilities in the basic unit.

Micron (m): A unit of measure, 10^{-6} m, used to measure wavelength of light.

Microprocessor: The control unit of a microcomputer that contains the logical elements for manipulating and performing arithmetical and logical operations on information.

Microscope, fiber optic inspection: A microscope used to inspect the end surface of a connector for flaws or contamination or a fiber for cleave quality.

Microwave: (1) Portion of the electromagnetic spectrum above about 760 MHz. (2) High-frequency transmission signals and equipment that employ microwave frequencies, including line-of- sight open-air microwave transmission and satellite communications.

MIPS: Million instructions per second. A measure of the speed of a CPU.

Misalignment loss: The loss of power resulting from angular misalignment, lateral displacement, and end separation.

Modal dispersion: Dispersion resulting from the different transit lengths of different propagating modes in a multimode optical fiber.

Mode: A single electromagnetic field pattern that travels in fiber.

Mode filter: A device that removes optical power in higher-order modes in fiber.

Mode: In guided-wave propagation, such as through a waveguide or optical fiber, a distribution of electromagnetic energy that satisfies Maxwell's equations and boundary conditions. Loosely, a possible path followed by light rays.

Modem: A device which connects a computer to a telephone line and sends data over those phone lines, normally using a modulated audio tone.

GLOSSARY *(cont.)*

Modem: A device which modulates and demodulates signals. It provides an interface between digital terminals and analog circuits and equipment.

Modulation: The process by which the characteristic of one wave (the carrier) is modified by another wave (the signal). Examples include amplitude modulation (AM), frequency modulation (FM), and pulse-coded modulation (PCM).

Module: The smallest replaceable unit in a PV array. An integral encapsulated unit containing a number of PV cells.

Motherboard: A board containing a number of printed circuit sockets and serving as a backing.

Motherboard: The central card of a computer that accepts other printed circuit cards. So called due to its female connectors and hierarchy in the computer. IBM now calls this a planar board in their System/2 computers.

Mouse: A handheld input device, separate from a keyboard, that is moved on a surface to control the position of an indicator (cursor) on a display screen.

Multi-access: The capability that lets multiple users simultaneously communicate with a computer or a network.

Multimode fiber: A fiber with core diameter much larger than the wavelength of light transmitted that allows many modes of light to propagate. Commonly used with LED sources for lower-speed, short-distance links.

Multiplex: To combine multiple input signals into one for transmission over a single high-speed channel. Two methods are used: (1) frequency division, and (2) time division.

Multiplexer: A device that lets more than one signal be sent simultaneously over one physical circuit. At the receiving end, the circuit is demultiplexed into the same number of outputs. Also called a mux.

Multiplexing: The process by which two or more signals are transmitted over a single communications channel. Examples include time-division multiplexing and wavelength-division multiplexing.

Multitasking: Two or more program segments running in a computer at the same time.

MUX: See Multiplexer

NA: Numerical aperture.

Nanometer (nm): A unit of measure, 10^{-9} m, used to measure the wavelength of light.

Near End Crosstalk (NEXT): Interference that transfers from the transmit side of a circuit to the receive side of the same circuit. It is measured at the transmit or "near" end.

NEC: National Electrical Code, which contains safety guidelines for all types of electrical installations.

GLOSSARY *(cont.)*

Netiquette: A pun on "etiquette"; proper behavior on a specific network.

Network: A computer network is a data communications system which interconnects computer systems at various different sites. A network may include any combination of LANs, MANs or WANs.

Network File Transfer: A procedure that lets a network user (1) copy a remote file, (2) /translate file attributes, and (3) access remote accounts, either interactively or through a program.

Network Meltdown: A state of complete network overload. The network equivalent of gridlock. See also broadcast storm.

Network Operating System (NOS): The software that manages the relationships between network resources and users. While there are several parts and protocols in a NOS, it is usually available as a single product.

Network Operations Center (NOC): The site for monitoring the operation of and directing the management and maintenance of a network.

Network Trunks: Circuits connecting switching centers.

Next: See Near End Crosstalk.

Node: An addressable device attached to a computer network. See also host, router, server, user.

NOS: See Network Operating System

Novell: Makers of NetWare software for networks and the dominant company in local area networking.

Numerical aperture: The "light-gathering ability" of a fiber, defining the maximum angle to the fiber axis at which light will be accepted and propagated through the fiber. NA**:** sin O, where O is the acceptance angle. NA is also used to describe the angular spread of light from a central axis, as in exiting a fiber, emitting from a source, or entering a detector.

OCR: Optical Character Recognition.

OEM: See Original equipment manufacturer.

Ohm: The unit of measurement of electrical resistance. One ohm of resistance will allow one ampere of current to flow through a pressure of one volt.

Open Circuit Voltage: The maximum voltage produced by a photovoltaic cell, module, or array without a load applied.

Optical amplifier: A device that amplifies light without converting it to an electrical signal.

Optical loss test set (OLTS): A measurement instrument for optical loss that includes both a meter and source.

GLOSSARY *(cont.)*

Optical power: The amount of radiant energy per unit time, expressed in linear units of Watts or on a logarithmic scale, in dBm (where 0 dB = 1 mW) or dB (where 0 dB = 1 W).

Optical switch: A device that routes an optical signal from one or more input ports to one or more output ports.

Optical time-domain reflectometry: A method of evaluating optical fibers based on detecting backscattered (reflected) light. Used to measure fiber attenuation, evaluate splice and connector joints, and locate faults.

Optical Character Recognition (OCR): a process than scans images, detects character shape matches, and converts them into digital code for the matching character.

Optical Disks: Storage devices that store data by using laser technology to record data. They Feature greater storage capacity than magnetic disks but currently offer slower access.

OS/2: Operating System/2. Graphical successor to DOS for the IBM PC. It was developed jointly by Microsoft and IBM but now sold only by IBM.

OSI Reference Model: A seven-layer structure designed to describe computer network architectures and the way that data passes through them.

OTDR: Optical time-domain reflectometry.

Out-of-band: Any frequency separate from the band being used for data, voice or video traffic. Typically requires a completely separate signal path or wire. Contrast with in-band.

Outlet: The place in the wiring system where the current is taken to supply equipment.

Overcurrent: Too much current.

PABX: Private Automatic Branch Exchange

Packet Switch: A device to accept, route and forward packets in a packet switched network.

Packet Switching: A communications paradigm in which packets (messages) are individually routed between hosts, with no previously established communication path.

Parity: Having a constant state or equal value. Parity checking is one of the oldest error checking techniques.

Parity Bit: A check bit appended to an array of binary digits for parity checking.

PBX: Private branch exchange. A telephone switch which is installed at the customer premises.

PC: See Personal Computer.

PCS: Plastic-clad silica.

Personal Computer (PC): (1) A desktop computer developed by IBM or a clone developed by a third-party vendor. (2) Sometimes used more generically to refer to other desktop systems, such as the Apple Macintosh. (3) The original, IBM developed, computer using an Intel 8088 cpu and an 8 bit internal bus.

GLOSSARY *(cont.)*

Phase converter: A device that derives three phase power from single phase power. Used extensively is areas (often rural areas) where only single phase power is available, to run three phase equipment.

Photodector: An optoelectronic transducer, such as a pin photodiode or avalanche photodiode.

Photodiode: A semiconductor diode that produces current in response to incident optical power and used as a detector in fiber optics.

Photon: A quantum of electromagnetic energy. A "particle" of light.

Photovoltaic: Changing light into electricity.

Physical Layer: the OSI layer (layer 1) specification that defines signal voltages, encoding schemes and physical connections for sending bits across a physical media.

Physical Media: Any means in the physical world for transferring signals between systems. Since it is outside (below) the OSI Model, it is also know as "Layer 0". It includes all copper media, fiber optic media and wireless technologies.

Physical Unit (PU): An SNA term used to refer to different types of hardware in the network.

Pigtail: A short length of fiber permanently attached to a component, such as a source, detector, or coupler.

Pixel: Picture Element. A single dot on a screen, or printout, that is represented by, or represents, a specific memory address.

Plain Old Telephone Service (POTS): Standard, two-wire, telephone service. Contrast with ISDN, Call Waiting, or any other additional features.

Plastic fiber: An optical fiber having a plastic core and plastic cladding.

Plastic-clad silica fiber: An optical fiber having a glass core and plastic cladding.

Plenum cable: A cable whose flammability and smoke characteristics allow it to be routed in a plenum area without being enclosed in a conduit.

Plenum: A chamber which forms part of a building's air distribution system, to which connect one or more ducts. Frequently areas over suspended ceilings or under raised floors are used as plenums.

Point to Point: A network configuration that has a connection between only two, terminal installations as opposed to multipoint.

Point-to-Point Protocol (PPP): A protocol for transmitting packets over serial, synchronous and asynchronous point-to-point circuits.

Polling: The process of inviting another station or node to transmit data. It requires that one station control the other stations.

POTS: See Plain Old Telephone Service.

GLOSSARY *(cont.)*

Power budget: The difference (in dB) between the transmitted optical power (in dBm) and the receiver sensitivity (in dBm).

Power meter, fiber optic: An instrument that measures optical power emanating from the end of a fiber.

Power (Watts): A basic unit of electrical energy, measured in watts. See also Watts.

Premises Distribution System: An AT & T cabling system that has also become a generic term for any structured cabling system.

Private Branch Exchange (PBX): A private phone system (switch) that connects to the public telephone network and offers in-house connectivity. To reach an outside line, user must dial a digit like 8 or 9.

Private Line (PL): A full-time, leased line that connects two points.

Process (verb): To perform one or more operations on information.

Process (noun): An operation performed on data or information.

Processor: A computer, or chip, capable of receiving information, manipulating it, and supplying results.

Program: A group of instructions that direct a computer's tasks.

Programmer: A person who designs, writes, and tests computer programs.

PROM: Programmable Read-Only Memory. A chip-based information storage area that can be recorded by an operator but only erased through a physical process.

Prompting: Messages from a computer that give instructions to the user.

Propagation Velocity: The speed that a signal travels through a medium from source to target. Also called Velocity of Propagation.

Protocol: A formal description of message formats and the rules computers must follow to exchange those messages.

Protocol Converter: A program or device that translates protocol in both directions.

Public Domain (PD) : Intellectual property that is freely available to all people.

Pulse spreading: The dispersion of an optical signal with time as it propagates through an optical fiber.

Punch Down: A wire termination technique in which the wire is laid on a connector and "punched" into place with a special tool. This procedure strips that portion of the insulation to provide an excellent, long lasting contact.

Radio Frequency Interference (RFI): The unintentional transmission of radio signals. Computer equipment and wiring can both generate and receive RFI. See also electromagnetic interference.

GLOSSARY *(cont.)*

Radius (Radii, plural): The distance from the center of a circle to its outer edge.

RAM: See Random access memory.

Random Access Memory: Dynamic memory, sometimes known as main memory or core. See also DRAM.

RBOC: See Regional Bell Operating Company.

Receive cable: A known good fiber optic jumper cable attached to a power meter used for loss testing. This cable must be made of fiber and connectors of a matching type to the cables to be tested.

Redundancy: (1) The portion of the total information contained in a message that can be eliminated without loss of essential information. (2) Providing duplicate, devices to immediately take over the function of equipment that fails.

Regeneration: The process of receiving distorted signals and recreating them at the correct rate, amplitude, and pulse width.

Regional Bell Operating Company (RBOC): One of seven regional holding companies that were the result of the AT & T divestiture (breakup).

Remote Site: Site not serviced by an electrical utility grid.

Remote Access: The ability of network nodes or remote computers to gain access to a computer which is at a different location.

Repeater: A device that receives, amplifies (and perhaps reshapes), and retransmits a signal. It is used to boost signal levels when the distance between repeaters is so great that the received signal would otherwise be too attenuated to be properly received.

Resistance: The opposition to the flow of current in an electrical circuit.

Resonance: A condition in an electrical circuit, where the frequency of an externally applied force equals the natural tendency of the circuit.

Response Time: The amount of time elapsed between a generation of an inquiry at a system and receipt of a response at that same system.

RG-58: The coaxial cable used by Thin Ethernet (10 base 2). It has a 50 ohm impedance and so must use 50 ohm terminators.

RG-59: The coaxial cable, with 75 ohm impedance, used in cable TV and other video environments.

RG-62: the coaxial cable, with 93 ohm impedance, used by ARCNet and IBM 3270 terminal environments.

Ring network: A network topology in which terminals are connected in a point-to-point serial fashion in an unbroken circular configuration.

RISC: Reduced instruction set computer.

GLOSSARY *(cont.)*

Rise time: the time required for the leading edge of a pulse to rise from 10% to 90% of its amplitude; the time required for a component to produce such a result. "Turnon time." Sometimes measured between the 20% and 80% points.

RJ: Registered Jack

RJ11: A standard six conductor modular jack or plug that uses two to six conductors. Commonly used for telephones and some data communications.

RJ22: The standard four conductor modular jack that connects a telephone handset to its base unit.

RJ45: A standard eight conductor modular jack or plug that uses two to eight conductors. Replacing RJ11 for use with data communications and increasing use with telephones. The wire may be twisted or flat though flat will only work up to 19.2 Kbps.

ROM: Read-Only Memory

Root Directory: The top of a namespace or file system. Usually represented by a slash (\ in DOS, / in UNIX).

Rotary convertor: A type of phase convertor.

Route: (noun) The path that network traffic follows from its source host to its target host. (verb) To send a packet or frame of data through a network to its correct destination. In other words, what routers do.

RS232-C: An EIA physical interface standard for use between data communications equipment (DCE) and data terminal equipment (DTE). The current standard is RS-232-E

Satellite Relay: An active or passive repeater in geosynchronous orbit around the Earth which amplifies the signal it receives before transmitting it back to earth.

SCSI: See Small Computer Systems Interface. Pronounced "scuzz-ee".

Semiconductor: A material that has electrical characteristics somewhere in between those of conductors and insulators.

Sensitivity: For a fiber-optic receiver, the minimum optical power required to achieve a specified level of performance, such as a BER.

Separately derived system: A system whose power is derived (or taken) from a generator, transformer or convertor.

Serial (computer): Handling data sequentially, rather than simultaneously.

Service: Equipment and conductors that bring electricity from the supply system to the wiring system of the building being serviced.

Service drop: Overhead conductors from the last pole to the building being served.

Shelf Life: The period of time that a device can be stored and still retain a specified performance.

GLOSSARY *(cont.)*

Signal Circuit: An electrical circuit which supplies energy to one or more appliances which give a recognizable signal.

Signal to Noise Ratio: Ratio of the signal power to the noise power in a specified band, usually expressed in decibels.

Simplex: Simplex mode. Operation of a channel in one direction only with no capability of reversing.

Simplex cable: A term sometimes used for a single-fiber cable.

Simplex transmission: Transmission in one direction only.

Sine Wave: A waveform corresponding to a single-frequency, periodic oscillation, which can be shown as a function of amplitude against angle and in which the value of the curve at any point is a function of the sine of that angle.

Single-mode fiber: An optical fiber that supports only one mode of light propagation above the cutoff wavelength.

Small Computer Systems Interface (SCSI): A standard for a controller bus that connects disk drives and other devices to their controllers on a computer bus. It is typically used in small systems.

Snail Mail: A derogatory term referring to the US Postal Service speed as compared to electronic mail.

SNR: Signal-to-noise ratio.

Software (computer): Programming, especially problem-oriented.

Solenoid: An electromagnet with a moveable iron core.

Source: the light emitter, either an LED or laser diode, in a fiber-optic link.

Specific Gravity: The ratio of the weight of the solution to the weight of an equal volume of water at a specified temperature. Used as an indicator of battery state of charge.

Spectral width: A measure of the extent of a spectrum. For a source, the width of wavelengths contained in the output at one half of the wavelength of peak power. Typical spectral widths are 20 to 60 nm for an LED and 2 to 5 nm for a laser diode.

Splice, fusion or mechanical: A device that provides for a connection between two fibers, typically intended to be permanent.

Splice: An interconnection method for joining the ends of two optical fibers in a permanent or semipermanent fashion.

SRAM: Static RAM

Star coupler: A fiber-optic coupler in which power at any input port is distributed to all output ports.

GLOSSARY *(cont.)*

Star network: A network in which all terminals are connected through a single point, such as a star coupler.

Step-index fiber: An optical fiber, either multimode or single mode, in which the core refractive index is uniform throughout so that a sharp step in refractive index occurs at the core-to-cladding interface. It usually refers to a multi-mode fiber.

STM: Synchronous Transfer Mode

STP: Shielded Twisted Pair.

Strength member: the part of a fiber-optic cable composed of Kevlar aramid yarn, steel strands, or fiberglass filaments that increase the tensile strength of the cable.

Surface emitter LED: A LED that emits light perpendicular to the semiconductor chip. Most LEDs used in datacommunications are surface emitters.

Surge Capacity: The requirement of an invertor to tolerate a momentary current surge imposed by starting ac motors or transformers.

Switch: Telephone equipment used to interconnect lines and trunks.

Switched Line: A communication link that may vary the physical path with each usage.

Systems Network Architecture (SNA): A proprietary networking architecture used by IBM and IIBM-compatible mainframe computers. With its widespread use, SNA has become a de facto standard.

T1: A digital carrier facility for transmitting a single DS1 digital stream over two pairs of regular copper telephone wires at 1.544 Mbps. It has come to mean any 1.544 Mbps digital stream, regardless of what transmission medium.

T2: A digital carrier facility used to transmit a DS2 digital stream at 6.312 Mbps.

T3: A digital carrier facility used to transmit a DS3 digital stream at 44.746 megabits per second. It has come to mean any 44.746 Mbps digital stream, regardless of what transmission medium.

T4: A digital carrier facility used to transmit a DS4 digital stream at 273m bps.

Talkset, fiber optic: A communication device that allows conversation over unused fibers.

Tap loss: In a fiber-optic coupler, the ratio of power at the tap port to the power at the input port.

Tap port: In a coupler in which the splitting ratio between output ports is not equal, the output port containing the lesser power.

TDM: Time-division multiplexing.

Tee coupler: A three-port optical coupler.

GLOSSARY *(cont.)*

Teflon: The Dupont brand name for HDPE resin. This material is often used to comply with smoke resistance requirements of the National Electric Code. Plenum cable is then frequently Teflon cable.

Telco: Local telephone company.

Telecommunications: The transmission of voice and/or data through a medium by means of electrical impulses that includes all aspects of transmitting information.

Telecommunications Outlet: The end of the fixed cable system in the users work area. Usually a wall or floor jack.

Termination: Preparation of the end of a fiber to allow connection to another fiber or an active device, sometimes also called "connectorization".

Test cable: A short single-fiber jumper cable with connectors on both ends used for testing. This cable must be made of fiber and connectors of a matching type to the cables to be tested.

Test kit: A kit of fiber optic instruments, typically including a power meter, source, and test accessories, used for measuring loss and power.

Test source: A laser diode or LED used to inject an optical signal into fiber for testing loss of the fiber or other components.

Thermal protection: Refers to an electrical device which has inherent protection from overheating. Typically in the form of a bimetal strip which bends when heated to a certain point. When the bimetal strip is used as a part of appliance's circuitry, the circuit will open when the bimetal bends, breaking the circuit.

Thyristor: A family of switching semiconductor devices, including SCRs, triacs, and diacs.

Time Division Multiplexing: A system of multiplexing that allocates time slices to each input channel for carrying data over an aggregate circuit.

Token Bus: A network access mechanism and topology in which all stations attached to the bus listen for a token. Stations with data to send must have the token to transmit their data. Bus access is controlled by preassigned priority algorithms. It is most used by ARCNet.

Token Passing: A network access method in which the stations circulate a token. Stations with data to send must have the token to transmit their data. See also Token Ring and Token Bus.

Token Ring: A network access method and topology in which a token is passed from station to station in sequential order. Stations wishing to send data must wait for the token before transmitting data. In a token ring, the next logical station is also the next physical station on the ring.

Topology: A standard method of connecting systems on a network.

GLOSSARY *(cont.)*

Total internal reflection: Confinement of light into the core of a fiber by the reflection off the core-cladding boundary.

Tracking Array: A PV array that follows the daily path of the sun. This can mean one axis or two axis tracking.

Traffic: 1) Calls being sent and received over a communications network. 2) The packets that are sent on a data network.

Transceiver: Transmitter-receiver.

Transducer: A device for converting energy from one form to another, such as optical energy to electrical energy.

Transformer: A device which uses magnetic force to transfer electrical energy from one coil of wire to another. In the process, transformers can also change the voltage at which this electrical energy is transmitted.

Transmission: The electrical transfer of a signal, message or other form of data from one location to another.

Transmission Speed: The number of bits transmitted in a given period of time, usually expressed as Bits Per Second (bps).

Traveling wave tube: A UHF electron tube in which a wave traveling along a helix interacts with an electron beam traveling down the center of the helix.

Trunk: A telephone circuit that connects two switches.

Twisted Pair: A type of cable in which pairs of conductors are twisted together to produce certain electrical properties. See also shielded twisted pair and unshielded twisted pair.

Tx: Transmit

Underwriters Laboratories (UL): A non-profit organization that was established by the insurance industry to test devices, materials and systems for safety, not satisfactory operation. It has begun to set standards. Items that pass the tests are marked UL Approved.

Uninterrupted Power Supply (UPS): Designation of a power supply providing continuous uninterrupted service.

Uninterruptible Power Supply (UPS): A device that provides continuous power in case the main power source fails. It includes filtering that provides a high quality AC power signal. See Standby Power System.

UNIX: A multi-user operating system developed by Bell Laboratories.

Unshielded Twisted Pair (UTP): Cable that consists of two or more insulated conductors in which each pair of conductors are twisted around each other. There is no external protection and noise resistance comes solely from the twists. See also Category Cable.

Up Link: The connection from an earth station to a satellite.

GLOSSARY *(cont.)*

UPS: See Uninterruptible power supply.

Usenet: The thousands of topically named newsgroups, the computers which run them, and the people who read and submit Usenet news.

Utilization equipment: Equipment which uses electricity.

UTP: See Unshielded Twisted Pair

Velocity of Propagation: See Propagation Velocity.

Virus: A program that replicates itself, usually to the detriment of the system, by incorporating itself into other programs which are shared among computer systems.

Visual fault locator: A device that couples visible light into the fiber to allow visual tracing and testing of continuity. Some are bright enough to allow finding breaks in fiber through the cable jacket.

Voice Frequency (VF): Any of the frequencies in the band 300-3, 400 Hz which must be transmitted to reasonably reproduce the voice.

Voice Grade: An access line suitable for voice, low-speed data, or facsimile service.

Volatile Storage: Any memory that loses it contents when electrical power is removed.

Volt: The unit of measurement of electrical force. One volt will force one ampere of current to flow through a resistance of one ohm.

Voltage Drop: Voltage reduction due to wire resistance.

VSAT: Very Small Aperture Terminal. A smaller dish for satellite communication.

WAN: Wide area network

WATS: Wide Area Telephone Service.

Watt: The unit of measurement of electrical power or rate of work. One amp represents the amount of work that is done by one ampere of current at a pressure of one volt.

Waveform: Characteristic shape of an electrical current or signal. The ace output from an invertor.

Wavelength division multiplexing (WDM): A technique of sending signals of several different wavelengths of light into the fiber simultaneously.

Wavelength: The distance between the same two points on adjacent waves; the time required for a wave to complete a single cycle.

Wavelength-division multiplexing: A transmission technique by which separate optical channels, distinguished by wavelength, are multiplexed onto an optical fiber for transmission.

WDM: Wavelength-division multiplexing.

GLOSSARY *(cont.)*

Whips: A flexible assembly, usually of THHN conductors in flexible metal conduit with fittings, usually bringing power from a lighting outlet to a lighting fixture.

Wide Area Network (WAN): A network that covers a large geographic area. See also Local Area Network, Metropolitan Area Network.

Wide Area Telephone Service (WATS) - A telco service that lets a customer make calls to or from telephones on designated lines, with a discounted monthly charge based upon call volume. Typically synonymous with "800" service. This is not free calling as SOMEONE must pay for every call.

Wideband: A term applied to facilities that have bandwidths greater than those needed for one channel.

Wire Center: The physical structure that houses one or more central office switching systems.

Wiring Concentrator: See concentrator.

Workstation: A networked computing device with additional processing power and RAM. Often, a workstation runs an operating system such as UNIX or Windows or Apple's System 7 so several tasks can run simultaneously.

Worm: A computer program that replicates and distributes itself. Worms, as opposed to viruses, are meant to spread in network environments.

Xfer: Transfer

Xmodem: An eight bit, public domain error checking protocol.

Ymodem: A file transfer protocol based on CRC Xmodem. Ymodem has a 1024-byte packet size.

Zip: To compress a file or program. Usually done with PKZIP.

Zmodem: An error-correcting, full duplex, file transfer, data transmission protocol for copying files between computers. It is faster and better than Xmodem and Ymodem.

About The Author

Paul Rosenberg has an extensive background in the construction, data, electrical, HVAC and plumbing trades. He is a leading voice in the electrical industry with years of experience from an apprentice to a project manager. Paul has written for all of the leading electrical and low voltage industry magazines and has authored more than 30 books.

In addition, he wrote the first standard for the installation of optical cables (ANSI-NEIS-301) and was awarded a patent for a power transmission module. Paul currently serves as contributing editor for *Power Outlet Magazine*, teaches for Iowa State University and works as a consultant and expert witness in legal cases. He speaks occasionally at industry events.